Science with a Human Face

In honor of Roger Randall Revelle

Science with a Human Face

In Honor of ROGER RANDALL REVELLE

Edited by

Robert Dorfman

Peter P. Rogers

Library of Congress Cataloging-in-Publication Data

   Science with a human face : in honor of Roger Randall Revelle / edited by Robert
Dorfman, Peter Rogers.
         p.   cm.
      Includes biliographical references and index.
      ISBN 0-674-79483-4 (pbk. : alk. paper)
      1. Science.   2. Environmental sciences.   3. Human ecology.   4. Population.
5. Revelle, Roger, 1909–1991.
      I. Dorfman, Robert., 1916–   .   II. Rogers, Peter P., 1937–   .
      Q171.S3764   1997   97-20687
      500--dc21   CIP

July 1997

TABLE OF CONTENTS

ROGER RANDALL REVELLE

1909–1991

ACKNOWLEDGMENTS

This volume of essays dedicated to the memory of Roger Randall Revelle is part of a sequence of expressions of the affection and admiration that he inspired. The first of these expressions was the Memorial Symposium attended by hundreds of people in the LaJolla Presbyterian Church a few days after his death. His four children, his children-in-law and several grandchildren offered tributes at the service, as did Revelle's life-long friend and colleague, Walter Munk.

At Harvard, where Revelle had organized the Center for Population and Development Studies and directed its activities for ten years, Revelle's many friends and colleagues felt an urgent need to express their sorrow and their deep affection for Revelle. Lincoln Chen, who had become Director of the Center, and the two editors conceived and organized the two-day symposium in Revelle's honor where these papers were first presented, but the symposium could not actually have taken place without the willing exertions of Christopher Cahill, Colleen Murphy, Beth Taylor, and other members of the Center's staff on whom the endless chores of arranging, revising, verifying, retyping, etc. devolved. Nor could it have been accomplished without the generous support of the John D. and Catherine T. MacArthur Foundation, arranged through the good offices and inexhaustible patience of Carmen Barroso and Victor Rabinowitch of the Foundation's staff. Nor without the unflagging support and encouragement of Ellen Revelle (now Ellen Clark Revelle Eckis) and members of the Revelle family, especially Mary Paci and Carolyn and Gary Hufbauer. To all of them we extend our heartfelt thanks.

After the seminar at Harvard in October of 1992, preparation of the volume extended over five years. One of the peculiar difficulties of the enterprise was the scattering of topics imposed by Revelle's remarkable breadth of scientific interests, including oceanography, population science, meteorology, economic development and some other fields. Arranging for publication accordingly presented difficulties, which were overcome eventually through the efforts of Lincoln Chen.

Members of the Center for Population and Development Studies's staff, especially Sofia Agras, Christopher Cahill, Winifred Fitzgerald, Colleen Murphy, Laura Reichenbach, Sagari Singh, discharged the seemingly interminable series of tasks entailed in getting the book ready for publication, with remarkable good humor and efficiency.

The editors are anxious to express their appreciation of the exceptional patience, good-will, and cooperativeness displayed by the contributing authors during the protracted editorial process.

We are grateful also to Deborah Day, librarian of the Scripps Institute of Oceanography, for contributing essential bibliographic and other data.

Robert Dorfman
Peter Rogers

Memoir: HOW ROGER REVELLE BECAME INTERESTED IN POPULATION AND DEVELOPMENT PROBLEMS

By Jerome B. Wiesner

When I became Science Advisor to President Kennedy, I asked him if I could use some of the political slots that were available to appoint assistant secretaries for research and development (R & D) in the departments of the government that had substantial amounts of research and development. Kennedy said that it was fine with him but that I would have to persuade each of the cabinet officers individually. They all agreed, and I then had the job of finding somebody acceptable to each of them. Fortunately, Kennedy was a very popular president so it was easy to persuade very good scientists to join his administration. In particular, I was able to persuade Roger Revelle to take the post of Assistant Secretary for R & D for the Department of Interior where he worked with Stewart L. Udall. They made a very good team.

A few months later, the President told me about a dilemma he faced. Mohammad Ayub Khan, the President of Pakistan, was about to make a State visit to the United States, and the President was worried that he would ask for weapons. Kennedy was not willing to give him weapons, because it would mean he would have to make a like commitment to India, which would ultimately lead to an arms race. Kennedy asked if I "[knew] of something we [could] offer Mohammad Ayub Khan instead of weapons, which would be so important that he would gladly accept it instead?" I didn't know much about Pakistan, and had no idea whether or not there was anything that would be an appropriate substitute. Fortunately, I remembered that I was planning to see Dr. Abdus Salam, Ayub Khan's Science Advisor, at a meeting at MIT a few days later, so I told President Kennedy that I would see if Salam had any ideas.

When we met, I asked Salam whether there was anything that President Kennedy could offer the President of Pakistan that would be particularly important to their country. I don't think I told Salam that this was to be a substitute for a weapons request, but I may have. At that time, Pakistan was having difficulties with salinization of much of their agricultural land. Salam said they were losing about one million acres of land each year, and it was becoming harder and harder to feed the people of Pakistan. In fact, at that time the country was not able to feed itself, and the shortage of food had to be made up by purchases and gifts of rice from many other countries, including the United States. We talked a bit about what might be done to alleviate the problem. It seemed to us to be a very difficult problem, but not an impossible one.

That evening I called President Kennedy and told him what I had learned. I asked for his permission to discuss the problem with Roger Revelle to see what he thought could be done. I called Roger and he was very interested in the challenge. This was Roger's normal reaction to a problem: the more difficult it was, the more interesting Roger found it. It was quite acceptable for Presidential Assistants like me to use Air Force transportation, so early the next morning, Salam and I went to Washington to meet with Roger. After a few hours of discussion it was clear that with enough pumping, which the Pakistanis were already doing on a small scale, it should be possible to lower the level of the aquifer, and to stop the salting of the soil. We even made some preliminary estimates as to how many millions of dollars the project would cost. In any event, it seemed to me that we had lucked out, because the problem that Salam told us about was so serious that if Kennedy proposed that we help with this dilemma instead of agreeing to Khan's request for arms, Khan might be quite satisfied.

The two presidents had a very satisfactory meeting, and the net result was that Roger Revelle and Abdus Salam started working together to alleviate the "water-logging" in the fertile fields of Pakistan.

Roger's first task was to assemble an expert team who knew about agricultural irrigation problems. The team, which was composed of scientists from Harvard, included Drs. Harold A. Thomas of the Division of Applied Sciences, Ayers Brinser of the Harvard Forest, and Robert Dorfman of the Economics Department (who is still there). Contributions were also made by the U.S. Geological Survey, specifically Thomas Maddock, an engineer. I recall an impressive geologist named James Isaacs of the Scripps Institution. There were also many experts from the Department of Interior's Water Resources Laboratories and Geological Survey, and other leading scientific organizations. While Roger was bringing his group together, Abdus went back to Pakistan and assembled a briefing team of Pakistani experts to transfer all of their existing knowledge to Roger's team when it arrived in Pakistan.

The Harvard team did some modeling of aquifers with computers, so for the first time it was possible to show with a mathematical model what the effect of different spacings and well sizes would be. It turned out that the present wells were too small and too far apart—their only effect was close to the well, so that most of the land continued to be water-logged. With this information it was possible to design a tubewell system that had some hope of solving the problem.

While this work was going on, Roger had another idea. He decided to look at the total food problem, instead of just the water problem. It was apparent that a given amount of money using a combination of agricultural improvements, as well as tubewells, would provide more food. The agricultural practices in Pakistan were unbelievably primitive. The farmers did not use water efficiently, they did not sow seeds well, nor did they plow deeply enough. Even worse, the farmers had no access to fertilizers or pesticides, and little thought had been given to the proper seeds for the climate and soil.

We went to Pakistan with a charter to investigate water-logging and salinization. It took Roger something less than two days to understand (before anyone else did) that that was not the basic problem. The real problem was thoroughly incompetent agriculture. The problem with water was the Irrigation Department, which was comprised of arrogant individuals who were more interested in cumshaw and in keeping the distributaries free of weeds than in delivering water when the farmers needed it. The uninformed workers from the Pakistani Agriculture Department were nearly as bad. Deliveries of fertilizer and other necessary supplies often arrived after the crops had wilted, if at all.

These were the conditions that deflected Roger's attention from the technical problems of water use (which would not confront the basic problems) to his visionary million-hectare development schemes. In the end, his great schemes proved too difficult; they were never implemented, and that incompetent government could never have carried them out. But his diagnosis was sound, and led to the brilliantly successful implementation of the many-faceted green revolution in the Punjab. Roger virtually single-handedly led the way out of the horrible, though heretofore unrecognized crisis of Pakistani agriculture. It was a triumph.

After very considerable study and discussion, Roger and his group decided to use a saturation technique employing the best agricultural methods that could be made to work at the hands of illiterate farmers. They planned to do this on one million acre plots of land, one per year for as long as needed. The plan called for providing fertilizers and pesticides, roads to make it possible to move produce to market, tubewells when necessary, and some simple education so that the farmers would use their water, fertilizer and seeds more effectively, and would also learn how to store their grain safely from pests.

I watched this exciting project from time to time, as the team of Revelle and Salam, supported by all the help the two presidents could give, moved into high gear. While the program was still in its early stages, President Kennedy was assassinated. I dealt with the shock and grief of the President's death, returned to MIT, and lost track of the program.

Many years later, I met Dr. Salam at a Third World Conference at the United Nations, and he told the group how Roger Revelle's insights had changed Pakistani agriculture from a grain importing economy to one which had surpluses to export—something I hadn't realized. This led to a discussion of the possibility of doing something similar in the Sahel in Africa, and after the meeting was over, the two of us approached Roger Revelle again, perhaps twenty years after his first triumph, and asked him if he was willing to try to do something similar to help the people of the Sahel. He said yes readily, and with the help of the Academy of Science, he assembled a group of people to explore the needs in that area.

The task was much more difficult than it had been in Pakistan because many countries in the area did not communicate with each other, due to a combination of Cold War animosities and local tribal conflicts. In addition to the difficult water and agricultural problems, the political problems made it almost impossible to see how to proceed. But, in spite of this, the studies began. One of the by-products of the first study was the creation of the African Academy of Science, which provided hope that scientists in the various countries could find a way to work together in spite of the deadly politics they faced.

Before it was possible to complete the preliminary studies, Roger died, and as far as I know the project came to a halt. Before he died, he and I had had preliminary discussions about bringing together Russian and American scientists who had expertise on the problem to help the African Academy complete the preliminary review. It would be a wonderful memorial to Roger Revelle if we could find a way to finance such a collaborative study. Such a study, if it were as successful as the first venture, might prevent reoccurrence of the famines being experienced today, and halt the spreading desertification, which would be a fitting tribute indeed.

INTRODUCTION

We are very proud to publish this collection of essays honoring Roger Revelle, the founding director of the Harvard Center for Population Studies. The contributions in this volume not only reflect first class scholarship that parallels Roger Revelle's intellectual interests but they also remain faithful to the high quality that characterized everything Roger Revelle did in his professional life.

I first met Roger 20 years ago when he invited me to present a seminar at the Harvard Center at 9 Bow Street. Intensely interested in the food and population dynamics of the Indian subcontinent, Roger chaired my seminar on health and population research in Matlab, a rural area of Bangladesh. Afterwards, Roger invited me into his office and, using a blackboard, probed my data and conclusions for an additional hour. As a young scientist, I began to appreciate that Roger's intellectual curiosity had few boundaries. His penetrating insights compelled me to rethink my analysis and he was able to help me reflect upon and improve my own work. All of this was accompanied by Roger's easy manner, never hinting at the huge gap of intellect and experience between two scientists at vastly different stages of professional life.

After I was appointed director of the Harvard Center in 1988, I immediately sought out Roger. Much to my relief, he seemed genuinely pleased over my selection, due to the value I think that he accorded to field experience overseas. After taking stock of the Center and its history, my admiration for Roger increased considerably, for I recognized that Roger, in launching the agenda of our Harvard Center, was decades ahead of his time. In the 1960s, nearly all American population centers focused on how to reduce rapid population growth, an undoubtedly important challenge. Roger saw, however, a broader agenda, then only a glimmering in the horizon. With growing populations, how would the world feed and care for the inevitably larger numbers of people, especially in the poorest regions of the world? Only by the 1990s did the interaction between population, the environment, and the quality of human life gain widespread intellectual salience. Revelle's intellectual leadership established a distinctiveness at the Harvard Center that his successors, including myself, have built upon.

In these and other interactions with Roger, I began to understand why he inspired such fierce loyalty among his colleagues and students. Some of that loyalty, affection, and love is reflected in the essays in this volume. Special thanks are due to Bob Dorfman and Peter Rogers for their initiative in organizing and bringing this volume to fruition. Our thanks also go to the Revelle family, especially Ellen Revelle and Mary Paci, who continue to express in many ways the Revelle commitment to health and well-being of all peoples around the world.

Lincoln Chen
October 18, 1996

I

EARTH SCIENCES

1

Slow Reactions and Old Seawater

By Edward D. Goldberg

Introduction

Roger Revelle's first scholarly activities in ocean science began with his doctoral work at the Scripps Institution of Oceanography. His early pursuits involved the carbonate system, especially as it related to the precipitation and dissolution of calcium carbonate in marine sedimentary cycles. He was deeply involved with the chemistry of carbon dioxide in sea waters— investigations which provided a springboard for his later work on the greenhouse effect and global warming.

Revelle approached oceanography in a dedicated and single-minded way. He was interested in understanding the marine environment for its own sake, and would draw upon conventional wisdom in chemistry, physics and biology to address his needs. As many of his colleagues know, he did this terribly well. But reciprocity with the basic sciences had little interest for him. He gave low priority to the general question of how studies in oceanography might contribute to understanding chemical, physical or biological processes. Yet there have been very rewarding two-way interrelationships between chemists and marine chemists for the past several hundred years. In honor of Roger, I would like to develop this interplay and argue that chemists of the ocean have impacted upon the mother discipline.

Chemists and the Oceans

Beginning with the birth of modern chemistry at the end of the eighteenth century, the ocean environment has been involved with chemistry's further development (Goldberg 1970). For example, the elements bromine and iodine were discovered in materials from the marine environment. More recently, the occurrence of two cosmic-ray-produced radionuclides was first found in the ocean domain: silicon-32 in sponges and beryllium-10 in sediments. Advances in analytical chemistry came about from the need to measure accurately very small amounts of marine chemical species. The assay of parts per billion levels of plant nutrients in seawater, such as phosphate and nitrate, were pioneered by marine analysts in the 1930s and 1940s and led to marked advances in microchemical techniques for dissolved substances.

The understanding of one ocean phenomenon played a partial role in the gaining of the Nobel Prize in Chemistry for Manfred Eigen in 1967. The story began during World War II when the U.S. Navy was concerned with the details of sound transmission in seawater. Sound absorption measurements in the ocean indicated anomalous increases (a factor of 30) in absorption beyond that due to fresh water at frequencies between 10 and 100 kHz. At the University of California at Los Angeles, Professor Robert W. Leonard and his students discovered, by making sound absorption measurements in solutions of major seawater components, that neither sodium chloride, magnesium chloride nor sodium sulfate exhibited significant absorption. However, a 0.014 molar solution of magnesium sulfate displayed the same kind of absorption observed at sea in propagation experiments.

On the basis of the extensive acoustic spectroscopy measurements in divalent sulfate solutions by G. Kurtze and K. Tamm at the Third Physical Institute in Goettingen, all made at atmospheric pressure, Eigen came up with a multistate dissociation model to explain the two relaxations observed. In particular, Eigen and Tamm hypothesized several quantitative models for magnesium sulfate. Of these models only one very closely corresponded to the pressure dependence of sound absorption and the pressure dependence of electrolytic conductance measured to over 1000 bars by Fred H. Fisher at the Scripps Institution of Oceanography.

Once magnesium sulphate had been identified as the culprit in the major processes of sound absorption in seawater, an explanation for the phenomenon was soon forthcoming. Magnesium sulphate stands out from the other salts in seawater through its high hydration, i.e., association with a large number of waters, and its strong effect upon viscosity. Eigen and his co-workers (Eigen and Tamm 1962) proposed a multistep association process for the interactions of magnesium ion, sulphate ion and water:

$$MgSO_4 \leftrightarrow Mg(O^H_H)SO_4 \leftrightarrow Mg\,(O^H_H)\,(O^H_H)\,SO_4 \leftrightarrow Mg^{++} + SO_4$$

These complex aggregates have the capability of absorbing a large variety of different types of energy. For example, the dissociation of the magnesium sulphate increases with increasing pressure, i.e., acoustic energy can be absorbed in this process.

Following the understanding provided by Eigen, novel models for the nature of seawater have been proposed. But more than this, basic chemistry has been enriched by these theories in understanding the structure and stabilities of salt solutions.

Old Seawater

The abyssal ocean may provide an unusual milieu to study very slow chemical processes inasmuch as its waters have been isolated for thousands of years from the atmosphere and from the sediments. One of the earliest measurements of extremely slow reactions involved the build-up of ferro-manganese minerals on the seafloor. These solid phases, composed primarily of iron and manganese oxides, are hosts to a variety of very rare metals including copper, cobalt, zinc, platinum, palladium and nickel, and have attracted a good deal of attention for potential future mining. On the basis of radiochemical dating techniques, the accumulation took place at a rate of 100 mm/million years for a precipitate from the Blake Plateau (Goldberg 1963). Subsequent studies have indicated rates of accumulation for other cases, expressed in a more descriptive way, between 1 and 100 atomic layers per day. These are indeed very slow reactions. To a laboratory chemist, fast reactions are studied with an oscilloscope where the reactants disappear or the products appear in billionths of a second or less. A slow reaction takes place overnight or several days at most.

Slow reactions taking place in seawater have not really been explored until recently. The deep waters have been out of contact with the atmosphere and the sediments of the seafloor for periods of thousands of years, based upon the carbon-14 ages of the dissolved organic species. There, extremely slow reactions must be found indirectly. Confusing such a search is the very nature of seawater itself—a very complex mixture of salts, waters and organic molecules. We can consider seawater as a sodium chloride solution with contributions of all of the members of the periodic table as well as a very large number of organic molecules. These entities can interact with each other, forming complex molecules whose persistence can be very short (seconds, minutes or even hours) and perhaps very long. We are well aware that chemical processes do take place in these waters. For example, dissolved organic matter is oxidized with overall disappearance rates of several thousand years. But the reactants and products remain obscure.

Somewhat by accident my laboratory has come across a phenomenon that may relate to old seawater providing a chemical system for the production of remarkably stable dissolved species that form over hundreds of years or longer. Our primary investigation involves the biogeochemistries of the platinum group elements, six metals in the center of the periodic table. They are situated in two horizontal rows, the second and third transition series. They are extremely rare in nature and are valued for their industrial and, more recently, medicinal uses. As a consequence of great difficulties in their analyses and low abundances in the environment, they have been studied only rarely. Yet one can systematically examine their behaviors in our surroundings, seeking coherence and dissimilarities—their so-called comparative chemistries (Goldberg 1984). Interpretations can provide insights into the environmental reactions of the metals. Comparisons within the platinum metal group or with other metals with similar chemistries can be rewarding.

TABLE I

Marine Chemical Abundances of the Platinum Group Elements

Element	Seawater pg/l Surface	Deep	Earth's Crust ppb	Deep-Sea Sediments ppb	Possible Seawater Redox States
Ru	2		0.4	1.0	III,IV,VII
Rh	50	100	0.4	0.60	III
Pd	20	60	4.0	3.2	II
Os	1.7		0.4	0.2	IV,VII,VIII
Ir	1.5		0.4	0.4	III,IV
Pt	100	250	4.0	3.8	II,IV

The marine distributions of the platinum-group elements are given in Table 1. Two elements, platinum and palladium, are an order of magnitude more abundant in crustal rocks than the other four. Yet in seawater, rhodium is amazingly the second most abundant element, falling just behind platinum. The deep sea sediment data more or less are in concord with the crustal rock data. Platinum is the most abundant element, rhodium one of the least. This is not unexpected as the sediments are, in part, formed from the resistant crustal minerals during the weathering cycle.

In seawater, rhodium is stabilized in solution, i.e., it is much less reactive than the other platinum group elements. The platinum and palladium ions form strong complexes with chloride, hydroxide and perhaps the amino acids. But what elevates rhodium to such a high concentration in seawater? A priori one would expect rhodium to be more reactive and hence, relatively more enriched in the solid phases than in seawater as is its vertical periodic table neighbor iridium. Of the six platinum metals, only rhodium has a valence state of three. Since it cannot be involved in redox reactions, a comparison of its chemistries with another element that only has a trivalent state seemed an appropriate strategy. Insights into their comparative chemistries can be most revealing. A comparison of the marine abundances of bismuth and rhodium are given in Table 2.

TABLE 2

Rhodium/Bismuth Weight Ratios in Compartments of the Marine Environment. Bismuth Seawater Values from Lee et al. (1985–1986)

	Rhodium/Bismuth
Pacific Ocean Waters	2
Pacific Ocean Macrophytes	0.1–5
Crustal Rocks (Levinson, 1974)	0.002
Coastal Marine Sediments	0.0004–0.0047
Ferro-manganese minerals	0.0006–0.003
Phosphorites	0.5

Noteworthy is the crustal abundance ratio of around 0.002 with bismuth three orders of magnitude greater than rhodium. The seawater concentrations are about double those of bismuth in both surface and deep waters, emphasizing the relative lack of reactivity of rhodium. The coastal marine sediments have values that are in the range of the crustal value, both higher and lower. This is not unexpected since the mineral compositions of the sediments, which contain the resistant unweathered minerals, vary to a large extent from place to place, with differing exposed surface rocks. A very substantial way to obtain the average composition of the earth's crust would be to bring together the coastal sediments from the continents of the world and to average their compositions.

Macrophytes, the large coastal algae, have ratios in Pacific waters that reflect the seawater average. Clearly, the algae accumulate these elements from the seawater itself and to a minor extent from particulate debris. There are systematic variations in the ratio that reflect differences in species and differences in locations. Similarly, phosphorites, which are formed from the dissolved species in seawater, have the seawater Rh/Bi ratios.

The similarity of the ferro-manganese mineral and coastal marine sediment ratios was unexpected. The high concentrations of bismuth of ferro-manganese minerals has been noted previously (Cronan 1976). The nodules accumulate these metals from the water itself. There is an unusual preference for the trivalent bismuth over the trivalent rhodium. This probably relates in part to the stabilizing factors involved with the dissolved forms of rhodium. There are some indications that the platinum group element concentrations in seawater may be determined by the relative kinetics of formation and breakdown of complexes, associations between positively charged ions and negatively charged ions and neutral molecules. From the viewpoint of chemists, rhodium and its vertical periodic table neighbors cobalt and iridium form complexes that are kinetically inert (Cotton and Wilkinson 1988), i.e., time constants of days. The long time constants are rather alien to chemists although Baes and Besmer (1976) do point out that the replacement of waters about the aquated rhodium ion by chloride ion has constants on the orders of months, short periods of time with respect to marine processes that can extend through millennia but long with respect to nearly all reactions in solution studied by laboratory chemists. Still, if complex formation in seawater can involve long times, an explanation for the unusually high Rh concentrations might be found. Li and Byrne (1990) attribute the higher concentrations of platinum relative to palladium in seawater to the very slow ligand exchange rates of Pt(II) relative to those of Pd(II). Whereas Pt has substantially higher seawater concentrations than Pd, their crustal rock abundances are similar.

Overview

Rhodium is the last stable element to be assayed in seawater. Its concentration is unexpectedly high. Conventional scientific wisdom does not suggest a thermodynamic explanation. I speculate that the kinetics of interaction of rhodium with one or more of the ligands in seawater can produce complex species with time constants of centuries or millennia. There are data in the literature illustrating that rhodium and other platinum metals can form sterically-hindered complexes with organic ligands where time constants exceed days and weeks. At the present time the identification of such complexes appears difficult with rhodium concentrations in seawater of the order of picomolar or less and the imprecise knowledge of potential ligands.

Deep-sea waters offer a unique system in which to study reactions in solution that take place over long time scales. I suspect that, as the details are formulated for the activities of elements in the ocean, the formations of complex species over long time scales may be identified as important factors.

In conclusion, as this paper has demonstrated, marine chemists have contributed novel concepts to chemistry itself through studies of the unique nature and antiquity of the seawaters. I suspect they will continue to do so in the future.

REFERENCES

Baes, C. F., and R. D. Besmer. *The Hydrolysis of Cations*. New York: John Wiley, 1976.

Cotton, F. A., and G. Wilkinson. *Advanced Inorganic Chemistry*. 5th ed. New York: John Wiley and Sons, 1988:908–12.

Cronan, P. S. "Manganese Nodules and Other Ferro-Manganeseoxide Deposits." In *Chemical Oceanography*, 2nd ed., vol. 5:217–63, edited by J. P. Riley and R. Chester. London: Academic Press, 1976.

Eigen, M., and K. Tamm. "Sound Absorption in Electrolyte Solutions due to Chemical Relaxation." *Z. Elektro Chem.* 62 (1962):107–21.

Goldberg, E. D. "The Oceans as a Chemical System. In *The Sea*, vol. 2:3–25, edited by M. N. Hill, 1963.

_____ . "Chemical Description of the Oceans." *Technol. Rev.* 72 (1970): 24–9.

_____ . "Comparative Environmental Chemistry of Metals and Metalloids." *Mar. Poll. Bull.* 15 (1984):281–84.

Lee, D. S., J. M. Edmond, and K. W. Bruland. "Bismuth in the Atlantic and North Pacific: A natural analogue to plutonium and lead." *Earth Planet. Sci. Lett.* 76 (1985–86):254–67.

Levinson, A. A. *Introduction to Exploration Geochemistry*. Wilmette, Illinois: Applied Publishing Ltd, 1974.

Li, J., and R. H., Byrne, "Amino Acid Complexation of Palladium in Seawater." *Environ. Sci. Techno.* 24 (1990):1038–41.

2

Global Ocean Warming

By Walter Munk

An Observation is Worth a Thousand Words

The realization that the burning of fossil fuel must be accompanied by an increase in atmospheric carbon dioxide (CO_2) and a consequent disturbance of the Earth's radiation balance goes back at least to Arrhenius (1896). The outstanding contribution made by Roger Revelle and Hans Suess (1957) was their proposal that this rate of increase should be measurable by techniques available even then, in 1956, and the steps they took that led to a demonstration of the proposal. The result was the twenty-five-year effort by David Keeling which is the starting point of almost all thinking on the subject.

It has been difficult to reconcile the clear-cut carbon dioxide (CO_2) time series with the temperature record of the atmosphere during the last hundred years. The oceans play a major role in the Earth's heat budget, and one needs to get the oceans right to get the atmosphere right. I am reminded of the situation Roger encountered in 1956. I believe that the heating of the oceans, if any, should be measurable by techniques available even now, in 1995.

The Oceans as a Sink of CO_2, Heat and Ignorance

There is a general consensus that the net production of carbon (allowing for burning of fossil fuels, clearing of forests and storage in the biosphere) exceeds the rate at which it is stored in the atmosphere by something like two gigatons per year. To close the budget it has been customary to store this excess in the oceans. It is only recently that a program has gotten underway to measure the variable CO_2 content of the oceans.

Similarly, the evidence suggests that the atmosphere is not accommodating to the increase in temperature required to maintain radiative balance, given the measured increase in the atmosphere of CO_2 and other greenhouse gases. About half of the increased surface flux, or 1 W/m^2, is available for ocean warming. This would correspond to a surface warming by 20 millidegrees per year (m^0C/y) in the surface mixed layer, decreasing exponentially to 5 m^0C/y at 1 km depth. Such an increase is consistent with a thermal expansion leading to a reasonable sea level raise by 2 mm/y. The Princeton and Hamburg coupled ocean-atmosphere models are in rough accord with this magnitude response to the CO_2 increase. (It has to be said at once that the ocean response is not uniform, but differs greatly with latitude and from basin to basin.) We shall take these numbers as a starting point.

Can it be Measured?

A group of Scripps, MIT and Woods Hole oceanographers have been working since 1978 on acoustic methods for ocean temperature changes. (A brief history of this work can be found in Munk et al. 1995). We propose an "acoustic thermometer" for measuring the ocean response to greenhouse warming. This unorthodox method is based on two considerations. First, the speed of sound is a function of temperature, increasing by about 4 m/s for each degree Kelvin. Second, the existence of an ocean sound channel (the SOFAR channel) makes it possible to retransmit over large distances, and this permits the measurement of large horizontal averages. Thus, the decrease in acoustic travel time between two distinct points is a measure of the increase in the average temperature along the transmission path.

The latter consideration is important, since local measurements are subject to large local variability. At a depth of 1 km, the depth of the sound channel, mesoscale eddies of typically 100 km scale are associated with month-to-month variations of order 1^0C, and this makes it impossible to detect a possible greenhouse warming by 0.005^0C per year. On the other hand, an average over 10,000 km sufficiently suppresses the mesoscale variability to make a detection possible. At this distance the projected greenhouse warming would lead to a decrease in acoustic travel time by something like 0.2 seconds per year.

A Global Sound Transmission

To achieve the precision required for measuring such changes requires electrically driven sources transmitting coded acoustic signals. A feasibility test was conducted in early 1991 to answer three questions: first, can existing electrically driven acoustic sources provide signals of adequate intensity at 10,000 km distances? Second, can the codes still be read? Third, can this be done without harmful effects to existing marine life? The answer appears to be yes on all three counts.

Adequate sources were provided by the U.S. Navy. These sources are limited to depths of less than 250 m, which made it necessary to go to high latitudes where the sound channel is relatively shallow. We chose a site near Heard Island in the southern Indian Ocean, which is "visible" (along great circles) from both the Atlantic and Pacific Oceans. The transmissions were clearly recorded in the North and South Atlantic, the North and South Pacific, the Indian Ocean and the Antarctic Ocean. We had planned to transmit for ten days, but a severe storm damaged all the equipment after five days. Roger sent us a message:

" To: Walter Munk   January 18, 1991
        Corey Chouest
        Walter,
Your messages from down under are wonderfully interesting.
Wish I was with you but glad I'm not.
                                Roger Revelle"

The Curious Whale

At a late stage, when acoustic receivers had already been shipped to eight participating countries, we were forcefully reminded of the unity of the marine sciences. The problem concerned possible harmful acoustic effects on marine mammals. It was not a question of causing physical damage, but rather of interfering with established behavioral patterns.

As a result, six American and three Australian biologists participated in the Heard Island Feasibility Test, and a second ship was chartered for the biological observations. The observers labored under difficult circumstances, as the experiment had not been planned with biological observations in mind. Aerial surveillance was ruled out; Heard Island is probably as far as you can get from a landing strip.

Our biological partners added greatly to the excitement and adventure of the experiment. We operated under a protocol that if any of the designated mammals were within sight of the source vessel at the scheduled transmission start times, such a transmission had to be aborted. Our principle fear was that a whale would be attracted into our vicinity by curiosity. As it turned out, there were no cancellations, and no evidence of interference.

Upon our return, we were disappointed to read (Cohen 1991) that "some" said we had harmed the whale population in the southern Indian Ocean (none of the participating observers had been consulted). In response, Roger wrote a letter to the Editor (Revelle 1991), which began, "Shame on you." That letter was to be his last published words.

One Man's Signal is Another Man's Noise

The chief difficulty in detecting the greenhouse signal in the ocean (or for that matter in the atmosphere) has to do with the large ambient variability. The oceans vary naturally on all space and time scales. (We have already mentioned a pronounced mesoscale variability.) The oceans also vary from decade to decade on a basin scale. How is one to distinguish the greenhouse signal from this natural variability of similar space and time scales?

Computer models indicate that even though the scales of ambient and greenhouse variability overlap, the structure of the variations is different. One needs to establish a global acoustic network of sufficient resolution to distinguish between the two. We are trying to specify what this means: how many stations recording for how many years are required to detect the greenhouse signal and probability p? Here we are leaning heavily on computer modeling of both ambient and greenhouse variations. The models may, of course, be way out of line. If they are, we believe that this is an argument for making measurements, not an argument against it.

The ambient ocean variability is of course a subject of great intrinsic interest. One man's signal is another man's noise. An acoustic network will contribute to our understanding of ambient variability.

What Next?

There are many remaining issues. We would like to record for at least one year from a fixed source to a fixed receiver to ascertain whether the signals are stable during a seasonal cycle. The Heard Island Feasibility test was necessarily limited to a few days of transmission, and the source ship was underway into the wind. (For the vessel to hove to would have been difficult if not dangerous.)

During the next two years we are planning for transmissions from Kauai northward to existing acoustic arrays at about 5000 km range, and from California southward to New Zealand (10,000 km). These transmissions are to be coordinated with critical observations of the behaviors of marine mammals in the source regions. Additional Japanese and Russian sources are under consideration. Source levels will be 20 dB below Heard Island intensity. We will optimize source and receiver design. The undertaking is necessarily international in scope. A SCOR (Scientific Committee for Ocean Research) working group has met twice, and we are more than encouraged by the interest shown by the international community. (Roger was a founder of SCOR.)

During these activities we plan to work with the modeling community to define the scope of required global networking.

Epilogue

To Roger Revelle, oceanography was a continuing adventure. He was quoted in Science (1973) as saying: "Oceanographers have more fun." He believed in an informed activism, with luck as an essential ingredient.

Roger's influence and friendship have played a crucial role in my entire career. Here I have mentioned a number of occasions when Roger's influence and interests intersected with my recent efforts in a most positive way. I must mention one more occasion. We were meeting in the office of the Administrator of the National Oceanic and Atmospheric Administration to attempt to find a solution to the problems posed by possible disturbance of marine mammals. Suddenly Roger appeared and sat down in a corner without saying a word. Walking was difficult and painful for him at that time, yet he had covered the endless corridors of Washington's Commerce Building to lend his quiet support.

References

Arrhenius, S. "Carbonic Acid in the Air Upon the Temperature of the Ground." *Phil. Magazine,* 41 (1896):237.

Cohen, J. "Was Underwater 'Shot' Harmful to Whales?" Science 252 (1991):912–14.

Munk, W., P. Worcester and C. Wunsch, *Ocean Acoustic Tomography*, Cambridge University Press, 1995.

Revelle R. R., and H. Suess. "Carbon Dioxide Exchange Between Atmosphere and Ocean and the Question of an Increase of Atmospheric CO_2 During the Past Decades." *Tellus 9*, no.1 (1957):18–27.

Revelle, R. R. "Do oceanographers have more fun?" Science 181 (1973):926.

_____ . Letters to the Editor. *Science* 253 (1991).

II

ENVIRONMENT AND DEVELOPMENT

3

ON SUSTAINABILITY

Robert Dorfman[1]

I first met Roger Revelle in September 1961 in the Hotel Oberoi in Lahore, Pakistan. That meeting was a turning point in my professional development and, I believe, in Roger's also, though in a very different way.

The meeting was arranged by two presidents who, I think, had never seen either of us. The inception was in the summer of 1961 when President Ayub Khan of Pakistan paid a state visit to President John F. Kennedy of the United States. As was customary during ceremonial visits by leaders of strategically significant countries, Kennedy asked Ayub whether there was anything he could do for him. Ayub replied, in effect, "I appreciate your asking. We are having a terrible time in the Punjab. Waterlogging and salinity are destroying the soil fertility there, and we are in urgent need of advice on how to save it." In response, Kennedy promised to send a team of experts to Pakistan to analyze the problem and prescribe a cure.

Kennedy fulfilled this promise by turning the problem over to his Science Adviser, Jerome Wiesner. A "White House/Interior Panel on Waterlogging and Salinity in West Pakistan" was established, Roger Revelle was recruited to be its chairman (although Revelle was an oceanographer and the affected area was 800 miles from the nearest ocean), and Harvey Brooks was asked to help find an economist. Harvey may be able to explain why an economist was included, but I think the reason was that the Panel was going to be large—about two dozen "experts" of one sort or another—and someone believed that it might be well to include an economist since the recommendations were likely to have economic consequences.

At any rate, Harvey called me up in the middle of August and invited me to serve as the Panel's economist. After checking an atlas to find out where Pakistan was, and looking in the *World Almanac* to find out what it was, I accepted and, a few weeks later, suffering from acute travel fatigue aggravated by jet lag, I climbed out of a taxi at the Hotel Oberoi, Lahore, a nostalgic, though seedy, relic of the British Raj. It was around cocktail time, and an invitation from Revelle was waiting for me, asking me to join him in his room. I delayed just long enough to wash my face and exchange my sweat-drenched shirt for a fresh one.

First impression of Revelle: an enormous man standing in a small room that exaggerated his bulk, with a welcoming smile on his face and a glass of bourbon in his hand. Second impression, after he had had a chance to greet me and begin to explain our mission: warm, unassuming, extremely quick-minded and insightful. Third impression, after a few days' acquaintance: a man who somehow appears cheerful and relaxed while applying his powerful mind and intense energy to the problem before him. Thirty years of acquaintance have proved those impressions to be somewhat limited but nonetheless unerringly accurate.

I started by saying that this meeting was a turning point for both of us. For me, it was my initiation into the problems of the less-developed countries and their need for social, political and economic development. Indeed, the project ignited in me an interest I have had ever since. For Revelle, the transition was very different.

He was already one of the most eminent natural scientists and scientific administrators in the country, having conducted highly influential oceano-graphic and atmospheric studies and having served as Director of the Scripps Institution of Oceanography during its decade of greatest growth. But some of his ambitions had been frustrated at the University of California and his interests were turning away from pure science and toward the application of science to social and economic policies. He had therefore accepted the post of Science Advisor to the Secretary of the Interior, a post that brought him to Washington and into the thick of those policy applications. The expedition to Pakistan introduced him to a new and critical range of science-policy questions: those confronting the less-developed nations. From then on, the center of his attention shifted from pure science and science administration to a deep concern with the great social problems of the mid-twentieth century, and particularly those of the impoverished Asian and African nations. He dedicated himself to using his scientific talents, skills and knowledge to solve, or at least ameliorate, the problems of those countries, and his guiding principle for himself and his colleagues came to be devoting their science to the service of humanity. He brought these interests with him three years later, when Jack Snyder, then Dean of the Harvard School of Public Health, attracted him to Harvard to establish the Center for Population Studies, and those interests became the guiding principle of the new center, which he directed for the next decade.

I

The phrase "sustainable development" had not yet been invented in those days. If it had, the Center might very appropriately have been called "The Center for Sustainable Development," for that was its major concern during Revelle's directorship. The Center's first major project was the completion of Revelle's study of Waterlogging and Salinity in West Pakistan, which centered on the interactions between the limited resources of the Punjab and the irrigation-based agricultural practices used there. From that time on, the interrelations among people, their society and economy, and the natural resources available to them, marked the basic style and concern of the Center.

Though the phrase "sustainable development" did not become fashionable while Revelle was actively engaged in development problems, it denotes the very issues that dominated the next twenty-odd years of his career. It also defines the motivating concept of the present paper. The closest thing there is to an authoritative definition was proposed by the World Commission on Environment and Development:

> Sustainable development is development that meets the needs of the present without compromising the ability of future generations to meet their own needs. It contains within it two key concepts:
>
> (1) the concept of "needs," in particular the essential needs of the world's poor, to which overriding priority should be given; and
>
> (2) the idea of limitations imposed by the state of technology and social organization on the environment's ability to meet present and future needs (World Commission 1987, 43).
>
> In essence, sustainable development is a process of change in which the exploitation of resources, the direction of investments, the orientation of technological development, and institutional change are all

in harmony and enhance both current and future potential to meet human needs and aspirations (World Commission 1987, 46).

As this definition indicates, there is an ineluctable tension between "sustainable" and "development." There are "limitations…on the environment's ability to meet present and future needs," but no guidance is given for reconciling the two demands. Instead, the definition ends on the inspiring note that in "sustainable development…the direction of investments, the orientation of technological development and institutional change are all in harmony and enhance both current and future potential to meet human needs…."

The definition propounded in the World Commission's report thus leaves open the question of whether sustainable development has ever been achieved or is even feasible. In 1991, Robert Solow (1991) proposed a more tractable concept that will serve as my point of departure. He shifted the frame of reference from positive description to normative demand. In this view, sustainability is seen as a moral obligation "to conduct ourselves so that we leave the future the option or the capacity to be as well off as we are…. Sustainability is an injunction not to satisfy ourselves by impoverishing our successors." And, more explicitly, "what we are obligated to leave behind is a generalized capacity to create well-being, not any particular thing or any particular natural resource." The emphasis on generalized capacity to create well-being is clearly derived from earlier work on natural resource economics by Dasgupta and Heal (1974) and Hartwick (1977), in which they studied the conditions in which a given stock of natural resources could support a prescribed level of consumers' welfare indefinitely. The principal objective of this paper is to extend this work to economies in which the "capacity to create well-being" is affected by the state of the environment as well as by the store of natural resources.

Hartwick (1977) discovered a remarkable rule which, under certain idealized, simplifying assumptions, would enable an economy to offset depletion of its natural resources and provide its consumers with a given level of welfare indefinitely, while its store of natural resources is being consumed. More specifically, Hartwick's rule holds that for an economy to be sustainable in the sense of being able to maintain its consumers' level of well-being indefinitely, it should invest neither more nor less than the net earnings, or hotelling rents, from consuming its natural resources in man-made fixed capital that will earn the same rate of return that the natural resources provided. These investments need not be related to the natural resources that are being depleted. Any adequately profitable investment in tangible capital or even human capital or knowledge will do. My reservations relate to the conclusion that such a broad range of possible investments could be adequate substitutes for environmental qualities.

Proofs of Hardwick's Rule invoke stringent assumptions. They include, among other things, constant population size, no technological change, no uncertainty about the future. In spite of these limitations of the formal analysis, it has been suggested that Hartwick's rule is the best "rule of thumb" available for guiding decisions that take responsible account of the welfare of citizens unto the third and the fourth generations and perhaps beyond (Solow 1986).

I have no quarrel with that. Hartwick's Rule is simple, intuitively satisfying, extremely ingenious and plausible. It is probably unrealistic to aspire to firmer or clearer guidance into the mists of the remote future. But I have strong doubts about extending the rule to include behavior with respect to deteriorating environmental conditions. Applying the rule to environmental deterioration would imply, for example, that increasing the stock of producible capital in a community, including the stock of gas-guzzling automobiles, could compensate for increased smog, or that the decimation of the redwood forests and their natural fauna and flora could be redressed by investing the net proceeds of the exploitation in unrelated but profitable man-made capital goods.

To be sure, it is difficult to distinguish between the natural resources for which Hartwick's rule can provide a useful rule of thumb and the environmental conditions for which, I maintain, it cannot. The dividing line that separates "the environment" from "natural resources" is smudged and, ultimately, unimportant. For present purposes a natural resource is a natural site or substance that can be owned and used privately with only negligible effects on the general public except, possibly, effects transmitted through market transactions. In contrast, the environment consists of sites and substances that are too directly connected to the general welfare to be entrusted to the discretion of private individuals or companies. In more technical language, the environment consists of sites and substances the use of which entails significant positive or negative externalities.

Typically, when a natural resource is used, it is consumed or depleted. In contrast, using an aspect or part of the environment generally has a negligible effect on the amount or quantity remaining, though it may impair or degrade its quality or usefulness. Indeed, this is the characteristic that I shall use to distinguish between natural resources and environmental conditions. Ordinary modes of use of natural resources reduce the amount that remains for later use; ordinary modes of using environmental conditions or features leave the amount essentially unchanged but affect their suitability for later use.

Since environmental conditions are neither created nor owned privately, and are not traded in normal economic markets, economists, until quite recently, have tended to take them as given—as free as air, one might say. It is, perhaps, a sign that the world is becoming crowded that we no longer use that simile. In that spirit, we shall say that the welfare of individuals and the productivities of firms depend not only on the private goods and services that they consume but also on the environmental conditions and public goods to which they are exposed willy-nilly.

In the sequel, we shall focus attention on the distinction between natural resources and the environment, and on the dependence of welfare on both by saying that an individual's well-being depends on only two things: c, a measure of the individual's consumption of ordinary goods, and E, a measure of the quality of the environment in which he, or she, or it (if it is a corporation) lives, thus suppressing many other ordinarily significant distinctions. Similarly, we shall divide the inputs employed by firms into only two classes: r, a measure of the complex of privately owned resources they use (including labor), and, again, E, the measure of the quality of their environments. Although such violent compression of the diversities of the goods used by individuals and firms, and of the components of the environments in which they operate, has some dangers, it has important advantages, including making it possible to portray the relationships among the environment, privately-owned resources, and production or consumption, largely in terms of two-dimensional diagrams.

II

In this section, we shall study the interrelations among the flow of consumable goods, changes in environmental quality, the use of resources to produce consumables both directly and by maintaining and increasing the stock of productive capital, and the use of resources to protect the environment and to restore it where necessary. The study will be conducted by means of three simplified, highly aggregated models. In the first model, consumers' well-being will depend on both the rate of flow of consumers' goods per capita and the quality of the environment, which is constantly being degraded by the same activities that produce the consumers' goods. In the second model, consumers' well-being depends only on the per capita flow of consumers' goods, but this output flow depends on environmental quality which, as in the first model, is affected adversely by productive activities. In the third model, both consumers' well-being and productive output are affected by the environmental quality, and environmental quality is affected by both productive activities and direct restorative efforts. We will find that only in the third model, where some resources are dedicated to preserving and restoring environmental quality, can non-declining levels of consumer well-being be maintained indefinitely.

Before considering the models, it will be helpful to introduce a conceptual device that figures in all of them. It is called an "indifference map" or "preference field," and depicts the psychological preferences of a typical consumer. Figure 1 is typical. It is a two-dimensional graph (though indifference maps of more dimensions are frequently used) in which the level of environmental quality, E, is measured horizontally and the rate of consumption per capita, c, is measured vertically. Each point in the figure therefore represents a particular c, E combination, to which a particular level of well-being, say $U(c,E)$ corresponds. The curves in the diagram are "indifference curves," on each of which, as we move from left to right, the advantages of increasing E are precisely offset by the drawbacks of decreasing c so that the consumer is completely indifferent among all the points on any one indifference curve. The curves are drawn to be smooth and concave-up (often called "convex"). The shape is important, because this shape, and only this shape, implies that the consumer will respond to relative price changes by shifting consumption gradually away from the good or goods that have become relatively more expensive.

There is also an intuitive explanation of the indifference curves' curvature. Consider a point near the northwest corner of the figure. It represents a situation with relatively high per capita consumption but low environmental quality. In such a situation, a marginal reduction in the level of consumption (e.g., forgoing a few luxuries) could be offset by a very small increase in environmental quality (e.g., a slight reduction in the frequency of smoggy days), as shown by the steep slope of the indifference curves in that region. Similarly, in the region where environmental quality is high but consumer goods are scanty, a small reduction in the provision of consumer goods would have to be offset by a marked improvement in environmental quality in order for consumers to remain on the same indifference curve, as indicated by the flatness of the indifference curves in the southeast part of the diagram. In general, indifference curves are drawn as convex to express the familiar fact that changes in kinds of consumption for which there is relatively skimpy provision are psychologically more significant than similar changes in kinds that are amply provided for.

If both high levels of consumption and high environmental quality are desirable, as we assume, the higher indifference curves in the diagram represent higher levels of well-being than the lower ones. The indifference diagram does not compare the amounts of satisfaction enjoyed on the different indifference curves since there is no scale for measuring degree of satisfaction.

Now we can present and interpret the three models.

Model I

In our first and simplest model, social well-being depends on both the rate of per capita consumption, c, and the level of environmental quality, E, but the economy's rate of output of goods and services per capita, to be denoted q, depends only on the stock of resources per capita, r. Thus, $q = f(r)$. We assume that $f(r)$ is a differentiable, smooth function of r and has a finite upper bound.

Dynamically, the model is driven by the decline in environmental quality resulting from productive activities, e.g., factory smoke or effluent discharge into public waters. This relation is expressed by

$$\dot{E} = -H(q),$$

in which the dot (˙) indicates a time-derivative and H(q) is the functional relation between the per capita rate of output and the rate of decline of environmental quality. We assume that H(q) is a non-negative, non-decreasing function and that it is greater than some positive number e whenever q > 0. By these assumptions, the environment deteriorates inexorably whenever there is any positive output. In order to sustain a constant level of social well-being, therefore, the rate of per capita consumption must continually increase.

In order for the rate of consumption to increase, the rate of resource use has to increase and so must the rate of total output per capita, the basic relationship connecting these magnitudes being

$$q = c + \dot{r} + Dr$$

where $\dot{r}$ is the rate of increase of per capita resources, or the rate of net investment, and D is a coefficient, assumed constant, expressing the rate of depreciation of a unit of resources per unit of time.

All the points on a single indifference curve correspond to the same level of well-being. Thus an indifference curve can be specified by a functional relationship such as $U(c, E) = $ constant. We will say that a given level of social

FIGURE I

Indifference Curves for Model I. Consumption Must Rise Increasingly to Compensate for Loss of Environmental Quality.

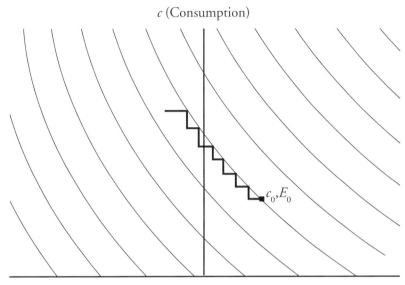

c (Consumption)

E (Environmental Quality)

well-being is sustainable if the social indifference curve[2] that corresponds to it can be attained and if the economy can move along it indefinitely.

Now we claim that under the assumptions of this model, no level of social well-being that requires a positive level of consumption is sustainable. A formal proof is presented in Appendix A, but the line of argument is indicated in Figure 1. A number of environment-consumption indifference curves are shown in the figure. The point (C_0, E_0) is an arbitrary consumption-environmental quality pair. On the staircase leading up and to the left from it, the treads (i.e., the decreases in quality) are all of equal length, but each riser (i.e., increase in the rate of consumption) is longer than the one before so that the steps can keep touching the indifference curve that contains (C_0, E_0). Eventually these steps will reach and exceed any ceiling, such as the upper bound to $f(r)$. At that point environmental quality will be so low that it will no longer be possible to produce the investment and consumer goods needed to sustain the prescribed level of well-being. The welfare level attained by (C_0, E_0) is therefore not sustainable. The same argument applies to any level of well-being.

The conclusion that no level of well-being is sustainable in this model results from the interaction of two assumptions. One is the assumption of a

FIGURE 2

Isoquants of the Production Function in Model Ii. Requirements for Resources Increase Increasingly as Environmental Quality Falls.

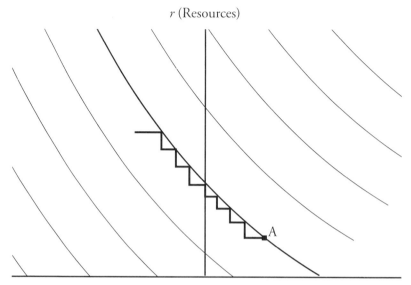

r (Resources)

E (Environmental Quality)

finite upper limit on the rate of production of consumption goods. The other is the assumption of concave social preferences, which requires an ever-increasing rate of output of consumer goods if the prescribed level of well-being is to be maintained while environmental quality falls. Both of these are stylized representations of characteristics of real economies that are difficult to deny.

Model II

The second model to be considered is, roughly speaking, a mirror image of the first. It assumes that environmental quality does not affect social well-being directly, but does affect the productivity of resources used in production. Since per capita consumption is the only factor affecting social well-being, it will be constant along any time-path on which a constant level of per capita consumption is maintained. As before, we shall assume that consumption or waste products associated with consumption harm the environment, and shall use the same kind of damage function as in the earlier model: $\dot{E} = -H(c)$.

The per capita production function for Model II is

$$q = f(r, E).$$

The dynamics of this model are simple. We have already stated the equation for the rate of change in environmental quality. The rate of change of output is the time derivative of the production function, or

$$\dot{q} = \frac{\delta f}{\delta r} \dot{r} + \frac{\delta f}{\delta E} \dot{E}$$

In this case, we shall say that a particular rate of consumption, c, is sustainable if it can be attained and if there is a time-path of net investment, $\dot{r}$, that enables that rate of consumption to be maintained indefinitely. No positive rate of consumption can be sustained for this model either, as can be seen by an argument much like the one used for Model I. The argument depends on the properties of differentiable concave functions, and can be indicated by an isoquant diagram (see Figure 2). In this diagram, levels of environmental quality are measured along the horizontal axis and available quantities per capita of manmade resources are shown vertically. The curves are "isoquants," each of which shows combinations of resources and environmental quality that can produce output at a particular rate. Now, consider any rate of output such as the one corresponding to the heavily emphasized isoquant, and some initial combination of resources and environmental quality that lies on the selected isoquant, such as point A. The staircase shows that ever-increasing amounts of manmade resources will be needed to offset the deterioration of the environment caused by the specified level of output. Sooner or later, therefore, the prescribed level of consumption will have to be abandoned. A formal proof is given in Appendix A.

Model III

In Model III, the principal substantive difference from the preceding two models is the introduction of the possibility of preventing, repairing or restoring damage to the environment. This change will make it possible for some positive levels of economic well-being to be "sustainable," meaning thereby they will be attainable and maintainable for an indefinite period. For a level of economic well-being to be sustainable, it must also correspond to an economic equilibrium, or else internal economic pressures will rapidly disrupt it. The model we have been working with must therefore be altered somewhat in order to yield an equilibrium. For simplicity we assume a stationary population and, what is worse, no technological progress, changes in tastes, etc. The resultant model, which resembles the preceding ones as closely as possible, consists of the following elements:

(1) A production function representing gross output per capita, q, as a function of environmental quality, E, and the amount of resources per capita used to produce commodities (as distinct from protecting or preserving the environment), r_1,

$$q = f(r_1, E). \qquad (1)$$

(2) An environmental protection function expressing the influence of the resources (per capita) devoted to protecting the environment, r_2, and the current state of the environment, E, on the rate at which the level of environmental quality is falling. This influence will be denoted by R, and the relationship by

$$R = g(r_2, E). \qquad (2)$$

(3) The contrary effect of production activity, q, and the current level of environmental quality, E, on the rate at which that level is falling. This relationship is expressed by

$$H = h(q, E). \qquad (3)$$

(4) The resultant of the last two relationships, which determines the rate of change of environmental quality:

$$\dot{E} = R - H \qquad (4)$$

(In the equilibrium that interests us $\dot{E} = 0$).

(5) The relation of the total stock of resources used per capita, $r_1 + r_2$, and the level of environmental quality, to the rate at which the stock of productive resources decays or depreciates:

$$d = D(r_1 + r_2, E). \qquad (5)$$

(6) The level of consumption per capita as determined by an aggregative demand-supply balance equation. In stationary equilibrium, the deterioration of the environment will be arrested, the capital stock will be constant, net investment will be zero, and the demand-supply balance will yield

$$c + pd = q, \qquad (6)$$

where p denotes the average cost of a unit of the capital good in terms of the consumption good.

FIGURE 3

Consumption Possibility Frontier and Welfare Indifference Curves for Model III.

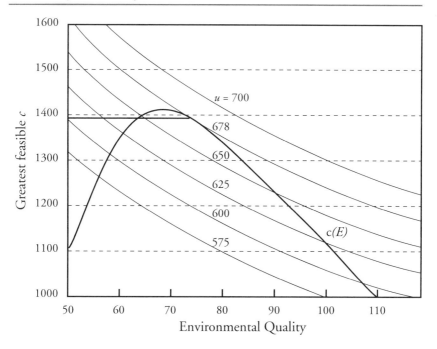

(7) The final requirement is that the configuration of choice variables, c, E, constitute a stationary equilibrium. That is, the solution must specify a level of environmental quality and provide as great a flow of consumables as is possible without either violating one of the constraints or reducing the environmental quality below the designated level. According to this provision, finding the equilibrium configuration of activity levels would amount formally to solving a maximization problem with inequality constraints, i.e., a Kuhn-Tucker problem, if a definite minimum acceptable level of environmental quality were prescribed. But that is not the case. Instead, choosing the level of environmental quality that permits the highest possible sustainable level of social-being to be reached is the heart of the problem.

The formal problem can be stated thusly: Consider the indifference map that describes the community's preferences among various levels of c and E. On it inscribe the curve, to be called $c(E)$, that shows the highest level of c consistent with maintaining each level of E over a relevant range. The point at which that curve touches, and is tangent to, the highest indifference curve that it reaches is the point where the variables satisfy all the conditions of the problem. Notice that the value of E at which this optimum is attained is not, in general, the value at which $c(E)$ reaches its maximum.

A numerical example of Model III is presented in Appendix A. This example may clarify the concepts and operations that have just been described, and also demonstrate that when repair and restoration of the environment are allowed, a sustainable equilibrium may be possible. Figure 3, which is based on this example, illustrates the configuration that results when the $c(E)$ curve is superimposed on an indifference map. This figure shows a number of community indifference curves, each of which represents a locus of combinations of average individual consumption, c, and environmental quality, E, all of which yield the same level of communal satisfaction. These levels are labeled, very arbitrarily, "575" up to "700." The curve $c(E)$ superimposed on the indifference curves show the highest level of c that can be attained in conjunction with each level of E.

The shape of $c(E)$, which necessarily portrays a special case, was chosen to indicate the variety of possible behaviors. In the left portion of the diagram, representing low levels of environmental quality, the quality of the environment is so low that increases in E permit increases in c. Thus the curve slopes upward as we move to the right. But as we move to the right, maintaining each unit increase in E requires more resources than its predecessor, and is less effective than its predecessor in increasing the efficiency of resources used to produce consumables. For both these reasons, $c(E)$ gradually becomes less steep until, when $E = 70$, the increase in efficiency is completely offset by the reduction in the amount of resources available to produce consumables. Thereafter, when E is increased still further, the increase in productive efficiency is overwhelmed by the need to divert resources from production to environmental preservation, so that the $c(E)$ falls, and falls ever more rapidly, as shown. Notice that, because of the slope of the indifference curves, $c(E)$ reaches the highest indifference curve it can attain at a level of E somewhat higher than the level at which per capita consumption is maximized. The optimum level of environmental quality in this example is about 78.

The example shows that, under proper conditions that conform to the standard assumptions of economic theory, a sustainable equilibrium can be maintained by devoting an appropriate share of the economy's resources to protecting and restoring the quality of the environment. Unfortunately, the example does not throw light on just what circumstances are required for such a sustainable equilibrium to be possible.

III

What, that wasn't obvious to begin with, have we learned from these three severely simplified models? I, at least, have learned a couple of useful lessons. I am most impressed by the finding that simply accumulating even large amounts of manmade resources does not necessarily compensate for environmental deterioration; specific attention to repairing or protecting the environment will generally be needed to sustain a desired level of consumer well-being indefinitely.

Dasgupta and Heal (1974) dealt with a somewhat similar problem in their study of exhaustible natural resources. They found that an economy could survive exhausting a resource if there were a renewable or producible resource whose elasticity of substitution for the exhaustible one was greater than unity.[3] Dasgupta and Heal's theorem clarifies the range of conditions in which Hartwick's rule is valid.

There may be a similar theorem applicable to situations in which consumers' well-being is affected by environmental conditions as well as by consumption of unreproducible natural resources, though, by definition, components of the environment are subject to being more or less degraded rather than being exhausted. If there is such a theorem, I haven't found it, though it may be lurking somewhere just below the surface.

The other main finding is simpler and more definitive: If there is some positive level of consumable output at which environmental deterioration can be arrested by using producible resources, then the corresponding levels of output, consumption and social welfare can be sustained indefinitely. It was easy to exhibit an example of a highly aggregated economy that satisfied normal economic hypotheses and which possessed such levels of consumption and resource use. This finding has no exact parallel in the theory of nonrenewable resources because it is not possible to continue to use a nonrenewable resource at a constant rate without using it up.

There are clear practical implications. By far the most important is a reaffirmation of the hoariest aphorism in economics: there's no such thing as a free lunch. Applied to the present context, that maxim holds that if sustainable development is a goal, a society will have to divert some investible resources from the conventional purpose of producing marketable goods and services to the protection and restoration of environmental quality. Prudent economic policies, then, have to admit lower levels of consumption and even of investment in the production of marketable goods in order to arrest, decelerate or even reverse the harm done to the environment by many economic and consumption activities.

Appendix A

This appendix establishes the principal properties of the three models discussed in the text, using arguments that are more formal than the ones given there. The notation is the same as in the text except when the contrary is stated.

Model I

We must show that the assumptions in the text imply that if any constant positive level of welfare is maintained, the rate of growth of the resources used in production, i.e. r, will increase without bounds. Since this is clearly impossible, there is no sustainable equilibrium with a positive level of output.

In this model, the level of social welfare depends on both per capita income, c, and environmental quality, E. We assume that each social indifference curve satisfies a relationship such as

$$u(c,E) = W \qquad\qquad (A1)$$

for some value of W, higher values corresponding to higher levels of social welfare. The function u(c,E) is assumed to be smooth, twice-differentiable, and convex, as explained in the text.

Suppose that some constant level of welfare, W, is sustainable. Total output must be sufficient to satisfy three components of demand: (1) demand for consumables, c, (2) demand for net investment to compensate for the inevitable decay of environmental quality, and (3) demand for resources to offset the effects of depreciation.

(1) The required rate of output of consumables per capita is a value of c that satisfies

$$u(c,E) = W.$$

As the quality of the environment falls over time, because $\dot{E} = -H(q)$, c will have to increase in order to satisfy equation (A1).

(2) The rate of decay of environmental quality is related to the rate of output, q, by $\dot{E} = -H(q)$.

To compensate for the environmental degradation, consumption must increase at the rate that maintains the level of welfare. By total differentiation of equation (A1), this rate is found to be a value, $\dot{c}$, that satisfies

$$u_1 \dot{c} + u_2 \dot{E} = u_1 \dot{c} - u_2 H(q) = 0$$

where u_1, u_2 denote partial derivatives of u with respect to consumption and environmental quality, respectively. Explicitly, then

$$\dot{c} = \frac{u_2}{u_1} H(q). \qquad (A2)$$

We will show in just a moment that $\dot{c} > 0$ implies $\dot{r} > 0$.

(3) By the simplifying assumption about depreciation, the effect of depreciation, which has to be offset, is just Dr, where r is the current level of resources.

Adding up these components, the rate of production, q, at any time must satisfy

$$q = c + r^{\cdot} + Dr.$$

Since q = $f(r)$, this can be written

$$f(r) = c + \dot{r} + Dr. \qquad (A3)$$

Equation (A3) is formally a first order differential equation. Whether it has an explicit closed form solution or not depends on the characteristics of the function $f(r)$; in general it doesn't have one. Fortunately, our argument does not require an explicit solution.

Now recall that the production function q = $f(r)$ signifies that q is the largest rate of output attainable with resources not greater than r, and, by the same token, r is the smallest rate of resource use that can attain an output of q. We maintain that equation (A3) requires that because c is always increasing and r must increase to accommodate it. If r did not have to increase, it would have been possible to attain the same level of consumption with a smaller input of resource services, contrary to the efficiency assumption.

Indeed, since, as we have argued, the set of c,E combinations for which $u(c, E) > W$ is convex, it is easily seen that the c,E indifference curves must also be convex (assuming that there is no satiation with respect to either c or E). Therefore, as time goes on and E falls, c must increase at an ever increasing rate. Specifically, as c and E move along a convex indifference curve,

$$\frac{d^2 c}{dE^2} = -\frac{1}{u_1^3} (u_{11} u_2^2 - 2 u_{12} u_1 u_2 + u_{22} u_1^2).$$

Since these partial derivatives are derived from a convex function, the quantity in the parenthesis is negative and $d^2 c/dE^2$ must be positive. This establishes that as E decreases along an indifference curve, dc/dE becomes steeper, i.e., the increases in c needed to compensate for unit decreases in E increase without bounds. This forces r and q to increase rapidly without limit, which contradicts the assumption that f(r) is bounded above. We must therefore abandon the assumption that the given level of welfare can be sustained.

Model II

The task in this model is to show that no positive level of consumption can be maintained indefinitely if consumption harms the environment in a way that reduces the productivity of resources used in production, and if nothing is done to prevent or repair the damage to the environment. The argument used to demonstrate this theorem is very similar to the one used to demonstrate the analogous theorem for Model I, which can be adapted to this model with little change.

The curvature properties ascribed to the production function, $f(r,E)$, are entirely conventional. They are that $f(r,E)$ is smooth, twice-differentiable, and concave, from which it follows that the isoquants are convex.

Suppose that a constant level of consumption, c, is sustainable. Total output must be sufficient to supply this level of consumption and, in addition to provide for the future by offsetting the depreciation of productive resources, Dr, and by sufficient net investment to compensate for the effect of environmental degradation on productive efficiency. The last component of demand for output is the only one that requires calculation.

The rate of environmental degradation depends on the level of consumption according to the relationship $\dot{E} = -H(c)$. The rate of net investment needed to offset the degradation will be denoted $\dot{r}$. The rate of output needed at any time to meet the three components of demand is then

$$f(r, E) = c + Dr + \dot{r}. \qquad (A4)$$

r is determined by the requirement that output keep abreast of demand. Taking the total derivatives with respect to time of both sides of equation (A4), and remembering that c remains constant, we obtain

$$f_1\dot{r} + f_2\dot{E} = f_1\dot{r} - f2H(c) = \frac{d\dot{r}}{dt} + D\dot{r}$$

This equation, together with $\dot{E} = -H(c)$, forms a pair of first-order differential equations that, in principle, determine the time-paths of r and E. Fortunately, since those equations need not be explicitly solvable, we do not need those solutions. All that is needed to prove our theorem is the assurance that $\dot{r}$ is always positive.

This assurance follows from the efficiency condition, i.e., the requirement that at all times r be the smallest level of resource services at which equation (A4) can be satisfied. Suppose there is a time-path of r along which it does not increase for some finite period. During that period, output, $f(r,E)$, must fall because E is decreasing at the steady rate $H(c)$. Therefore, since consumption remains constant by hypothesis, equation (A4) requires that r falls. At the end of the no-growth period, the stock of resources will be smaller than at the beginning. If the economy can continue to support consumption at rate c indefinitely, the initial stock of resources must have been unnecessarily large. This shows that an efficient time-path of quantities of resources cannot contain a horizontal or declining segment.

More than that can be said using the fact that $f(r,E)$ is concave. Then, just as with the welfare indicator in Model I, as E and r move along an isoquant the partial derivatives of $f(r,E)$ must satisfy

$$\frac{d^2 f}{dE^2} = -\frac{1}{f_1^3} (f_{11}f_2^2 - 2f_{12}f_1f_2 + f_{22}f_1^2) \geq 0.$$

We are not now concerned with movements along an isoquant but with somewhat steeper paths that do not fall below the isoquants on which they originate. Because the quadratic form in parentheses is negative and f_1 is positive, $d^2 r/dE^2$ must be positive. Therefore, as time goes by and E decreases steadily, r must increase more and more rapidly. But the rate of output, $f(r,E)$, cannot keep up with this accelerating demand. Instead, as r increases E has to decline toward zero fast enough to keep $f(r,E)$ at or below its finite upper bound. Therefore, consumption at the rate c with its accompanying decline of environmental quality at rate $H(c)$ cannot continue indefinitely, as was to be proved.

Model III

In the text, an algebraic instance of Model III was described, and it was alleged that numerical examples of that model which permitted a sustainable equilibrium could be constructed. The following numerical model is presented in support of that claim. The functional forms and numerical coefficients were chosen to conform to all the assumptions of the general case as presented in the text. Specifically, the equations in the numerical model are:

(1) Production function:

$$q = f(r_1,E) = 3r_1^{.6}E^{.3}.$$

(Recall that in this model r is split into two parts: r_1, the quantity of resources used for production, and r_2, the quantity of resources used for environmental protection and repair.)

(2) Environmental protection function:

$$R = g(r_2, E) = \frac{1}{\sqrt{E}}(205 - \frac{200}{1 + \dfrac{r_2}{265}}).$$

(3) Environmental damage function:

$$H = h(q, E) = \frac{q}{100}(\frac{E+25}{125})^{1.1}$$

(4) Resource depreciation or decay function:

$$q = f(r_1, E) = 3r_1^{.6}E^{.3}.$$

In addition, two equilibrium conditions

$$\dot{E} = R - H \geq 0,$$

and

$$\delta + c \leq q$$

have to be satisfied.

Subject to all these conditions, the task is to find a value of E and the corresponding value of c that make the value of $u(c, E)$ as great as possible. As stated in the text, the formal problem is to find a level of environmental quality, E, and a corresponding level of consumption, c, (not necessarily unique) such that the pair E and $c(E)$, the greatest sustained rate of consumption consistent with E, lie on the highest possible social indifference curve. The relevant social indifference map is taken to be the family of curves satisfying

$$c^{.75}E^{.25} = \text{constant.}$$

Figure 3 shows this indifference map together with the function $c(E)$ showing the greatest value of c for which the pair c, E can be sustained in light of the constraints imposed by the problem.

As explained in the text, the optimizing pair, c, E can be found in general by a two-stage computation. The first stage consists of selecting a number of values of E, and for each value solving the nonlinear programming problem of finding the greatest value of c for which all the constraints listed can be satisfied. This is $c(E)$. The second stage consists of finding a value of E for which $c(E)^{.75}E^{.25}$ is greatest. In the present instance, however, this tedious procedure is unnecessary. Since there are only two variables, the optimal pair can be found directly by using a standard optimization program. The solution is $c = 1391$, $E = 78.5$, as shown in Fig. 3.[4]

This example establishes that an economy that devotes some of its resources to repairing and protecting its environment can, in some instances, maintain a sustainable equilibrium. Unfortunately, it does not shed much light on the circumstances which are necessary for the realization of the possibility.

Appendix B
Definitions and a Theorem

This appendix will explain and discuss some unavoidable technicalities that are needed to justify the social indifference curves used in the text.

Private and common or public goods. Private goods, also called commodities, are goods that can be divided into quantities small enough to be appropriated, owned, and used by a single individual to the exclusion of all other individuals. Common or public goods are goods that cannot be so appropriated, for either technical or public policy reasons, but are used or experienced in common by the members of a community. Public goods may also be "bads" which members of a community must endure willy-nilly, for example, air with a high concentration of sulfur dioxide.

For our purposes, the essential distinction between public and private goods is that each individual chooses the quantity of private goods it uses or consumes, but has only a negligible influence on the quantities and qualities of the public goods that affect it for good or ill.

Individual preferences. The word "individual" is used to denote ordinary individuals and also groups of people who habitually make consumption decisions jointly and consume many commodities in common. It includes single individuals living alone, families, households and other noninstitutional living arrangements.

Individual states. An individual's state is a specification of its rates of consumption of all commodities, and of the amounts and conditions of the common goods in its community.

Note: Throughout this paper we assume that all individuals in a community are affected by the same common goods. That is an extreme simplification; individuals in California and Massachusetts are members of the same community for some purposes, but are exposed to quite different sets of common goods for the most part. This simplification facilitates the exposition without concealing or altering the principles under discussion.

Preferences, ranking of states. Each individual is assumed to be able to compare the members of any pair of states that are possible for it, and to decide which of the states it prefers or if it is indifferent between them. We also assume that these pairwise comparisons are transitive, that is, if the individual prefers state A to state B and B to C, then it prefers A to C. Thus, the individual's rankings form a "preordering" in mathematical terminology.

We also assume that every individual would prefer to consume more rather than less of every private good, and that public goods are specified in such a way that higher numerical ratings correspond to preferable characteristics. For example, since high concentrations of sulfur dioxide, particulates and other atmospheric pollutants are undesirable, numerical measures of concentrations of pollutants should be treated as negative numbers.

Indifference set or curve. A set of states that an individual regards as neither more nor less desirable than each other form an indifference set.

Superior set. The states that an individual regards as desirable as or more desirable than a given state form the superior set to that given state for that individual.

Key assumption. All of every individual's superior sets are closed and convex. The justification for this assumption is discussed in the text.

Social Welfare

Social state. A community's social state is a specification of the total rates of consumption of all the private goods consumed in it, plus the amounts and characteristics of the social goods or goods used in common in that community.

Pareto efficiency. A state is said to be Pareto efficient if there is no way to reallocate the private goods available in it so as to increase the satisfaction of any individual without reducing that of some other individual, or to increase the quantity or quality of any public good without decreasing the quantity or quality of some other public good or some private good. Only Pareto efficient states will be considered in the comparisons discussed below.

Ranking of Social States. Social State A will be said to be preferred to State B if it is Pareto efficient and if, with respect to every allocation of private consumable commodities possible in State B, there is some possible allocation in State A, that every individual regards as at least as satisfactory as the one in State B, and the reverse is not true.

Comment: The ranking of social states, unlike that of individual states, is not necessarily complete. That is, there may be no way to allocate the goods available in State A so that every individual regards itself as as well off as in some allocation attainable in State B, and simultaneously no allocation attainable in State B that everyone regards as as satisfactory as some allocation of goods available in State A. Such pairs of states are said to be incomparable, and if there are such pairs the ranking of social states is said to be incomplete.

Theorem. A ranking of social states is transitive.

Proof. This follows at once from the assumption that the rankings of individual states are transitive.

Social indifference set. A social indifference set is a set of comparable social states such that each individual is as well satisfied with his or her individual state in any of them as in any of the others.

Social superior set. All the social states that are at least as desirable to all individuals as some designated one constitute the social set superior to that one.

Main theorem. If all the individual superior sets in some community are closed and convex, so are all the social superior sets.

Proof of convexity. We first consider groups of states all of which have the same vector of public goods but differ with respect to their private goods vectors.

Consider a community in which all individuals' superior sets are closed and convex. Let State A and State B be two social states contained in the superior set to State C. Let ai denote the vector of private goods allotted to a typical individual, i, in state A and let bi be that individual's vector in state B. Since both A and B are in the superior set to C, individual i prefers both ai and bi to ci, its private goods vector in C. And since the individual superior sets are convex, they also prefer any internal linear combination of ai and bi, such as $\alpha a_i + (1 - \alpha)bi$, $0 \leq \alpha \leq 1$, to c_i. This being true for all individuals, state α is preferred by everyone to state C, and is in its superior set, which is therefore convex.

A similar argument applies if States A and B yield the same vector of private goods but different vectors of public goods. Then, if both A and B are contained in $\alpha A + (1 - \alpha)B$ the superior set to State C, every individual must at least weakly prefer both public goods vectors to State C's vector, and, since individuals' superior sets are convex, must also prefer any interior linear combination of A's and B's public goods vectors to C's vector. Thus, again, the superior set to State C is convex.

Finally, the same argument applies if the public goods vectors in States A and B also differ. Consider any individual i. If i prefers both States A and B to State C, then since its preferred set is convex, it must prefer any interior average of States A and B to State C. But these assumptions hold for all values of i, so the interior average lies in the preferred sets of all individuals, i.e., is in the set socially superior to C.

REFERENCES

Arrow, K. J. and A. Enthoven (1961), "Quasi-Concave Programming." *Econometrica* v. 29, 779–800.

Baumol, W. J. and D. F. Bradford (1972), "Detrimental externalities and non-convexity of the production set," *Economica*, v. 39, 160–176.

Chiang, A. C. (1974, 1984), *Fundamental Methods of Mathematical Economics.* New York: McGraw-Hill.

Dasgupta, P. and G. Heal (1974), "The optimal depletion of exhaustible resources," *Review of Economic Studies*, Symposium, 1974, 3–28.

Hartwick, J. M. (1977), "Intergenerational equity and the investing of rents from exhaustible resources," *American Economic Review*, v. 67, 972–974.

_____ . (1978), "Substitution among exhaustible resources and intergenerational equity." Review of Economic Studies, v. 45, 347–354.

Nordhaus, W. D. (1991), "To slow or not to slow: The economics of the greenhouse effect." *Economic Journal*, vol. 101, 920–937.

Revelle, R. R. (1974), "Will the earth's land and water resources be sufficient for future populations?" in United Nations, Department of Economic and Social Affairs, The.

Population Debate: Dimensions and Perspectives, Papers of the World Population Conference Bucharest, 1974. New York: United Nations.

Solow, R. M. (1986), "On the intergenerational allocation of natural resources," *Scandinavian Journal of Economics*, v. 88, 141–149.

_____ . (1991), "Sustainability, an economist's perspective." The Eighteenth S. Seward Johnson Lecture, Woods Hole Oceanographic Institution, June 14, 1991.

Starrett, D. (1972), "Fundamental non-convexities in the theory of externalities." *Journal of Economic Theory*, v. 4, 180–199.

The World Bank (1992), *World Development Report 1992, Development and the Environment.* Washington, DC: The World Bank.

World Commission on Environment and Development (1987), *Our Common Future.* New York: Oxford University Press.

NOTES

1. I want to express my debt to Ms. Sofia Agras and Professors John W. Pratt and Peter Rogers for their extremely constructive comments on earlier drafts of this paper as well as on the present version.
2. See Appendix B for the relationship between individual and social indifference maps, and related concepts.

3. "Elasticity of substitution" is a standard economic concept, but to define it here would take us too far afield to be attempted. Strictly speaking, Dasgupta and Heal's proof applies only to curves with constant elasticity of substitution, but it seems reasonable to believe that the conclusion applies approximately to other types of curves with similar shapes.
4. Chiang (1974, 1984), pp. 391–394 in the 2nd (1974) ed.
5. I thank Professor Christian Dufournaud of Waterloo University, Canada and Mr. Nagaraja Harshadeep of the Division of Applied Sciences, Harvard University, for programming and executing the solution to this problem.

4

America's Changing Environment Revisited Foresights, Hindsights, and Insights on Change

By Hans H. Landsberg[1]

The Setting

In the summer of 1966 Roger Revelle organized a conference at the Woods Hole house of the National Academy of Sciences at which seventeen scholars discussed and appraised trends and changes in the environment that they judged to merit attention as being imminent and, for the most part, ominous, if allowed to continue. Out of the conference emerged an issue of DAEDALUS (Fall 1967) and subsequently, with the addition of two more essays, a book (Revelle and Landsberg 1970) that, in essence, represents an early attempt to take stock of changes around us, as seen from the perspective of the mid-sixties. At Roger's invitation I became his coeditor and joint author of the Introduction.

Originally, the undertaking was a component of an ambitious Academy of Arts and Sciences project to identify tendencies in various segments of American life that might affect the next decade. A parallel effort that did not, however, come to full fruition, had as its aim to glance as far ahead as the year

2000. In contrast *America's Changing Environment* was not linked to a fixed target year. Rather its aim was to look more loosely at the "passing parade" without precise benchmarks.

It is instructive, sobering and sometimes embarrassing to compare the trends that these thoughtful specialists anticipated (or ignored), with what has actually happened. Some of the contrast between thinking in the mid-sixties and current perceptions is, I believe, due in part to the changing connotations of the term "environment." With gross simplification one might say that the earlier emphasis was on the immediate surroundings, the palpable, the obvious, and, by and large, natural, inherited or given environment. Gradually, attention has moved toward the larger systems that are the moving forces, the drivers that are the recipients of man-made impacts. In this paper, I try to identify these shifts in the meaning of the term "environment."

Traps and Biases

The world is too full of problems pressing for solutions to leave much time and space for returning to past studies and evaluating them in terms of current realities. Moreover, there is a temptation to operate from hindsight and, according to the evaluator's own biases, find fault with or admire the degree to which assessments and predictions came close to hitting the target. These are intriguing tasks well worth pursuing, but there is a great paucity of such enterprises. That is too bad, for something can be learned from them. Not only humility, but what sets the agenda at any given time, and how rapidly it can change.

Assessment is easiest when one deals with numbers. They are either right or wrong, or, more generously, either in the ballpark or not. I have occasionally indulged in such undertakings, looking back, in one instance, at the predictions of the Paley Commission (Resources for Freedom 1987), and in another at a portion of my own work, specifically energy forecasts made in the early 1960s (Landsberg 1960). I concluded then that one may identify a number of "traps" that contribute to biasing one's forecasts. Prominent among them is the tendency to become enamored of technological advances and ascribe near-term impact to them. Typically neglected are the economic and institutional obstacles that deter adoption of technological advances. Another trap is the difficulty of purging one's mind of normative thinking, in the sense that one tends to expect certain developments to take place, because they seem so obviously the "right thing" to do. I recall a specific instance, when in forecasting new housing in the U.S., my colleagues and I, in 1959 or thereabouts, were persuaded that the housing situation in the United States was intolerable and should, therefore, be remedied by a massive construction program (Landsberg, Fischman and Fisher 1963). This led to a great overestimation of everything connected with housing (e.g., lumber, steel, etc.), largely because factors such as interest rates and other shapers of housing activities had been ignored in the presence of imperative "needs." Yet another source of bias, most difficult to remedy, is the sheer inability to foresee social upheavals. The rapid growth of female participation in the labor force in the 1960s and 1970s is an example. It occurred unheralded and played havoc with labor force predictions and the macro variables related to it. So did the growth in single-person households. Others are changes in diet, though they generally will affect only segments of the economy, not society as a whole.

On a very large scale of the unpredictable, who would have ventured to suggest that within three decades or so of the near-destruction of the German and Japanese economies they would literally rise from the ashes and become potent competitors of the victors of World War II? Space travel, nuclear energy, computers, biotechnology and a host of other innovations happened without much advance notice, but with massive consequences.

For better or for worse, the essays that make up the bulk of *America's Changing Environment* are short on numbers, making it difficult or impossible to pass quantitative judgment on the findings and observations. That shifts the focus to (1) the selection of topics, (2) the yardsticks and criteria used to assess each segment, and (3) significant misconceptions. What follows is an attempt to address these topics, not exhaustively, but selectively in an illustrative manner.

The book does contain some attempts at quantification. The introductory remarks, especially, address the setting, and here numbers abound, although basically only for the purpose of providing "atmosphere." Tagging growth in population and GNP as the principal movers in the changing environment is supported by showing the changes that in consequence occurred between 1940 and 1965 in the number of automobiles, paper consumption, fertilizer use, and so on. But the Introduction largely refrains, wisely, I believe, from engaging in forecasts, though the unspoken implication is obvious: at a continuation of the 1940–65 rates of change we are courting trouble, if not disaster, especially because in addition to growth in population and GNP, human behavior is an important contributor to menacing environmental trends.

Energy: A Major No-Show

When it comes to the selection of topics, the most striking aspect of the collection is the absence of a separate chapter, or even of a segment in one of the other chapters, on energy. Today the tight linkage between energy and the environment is an undisputed facet of environmental research and policy-making, to the point where some analysts maintain that energy policy is in fact environmental policy, or put differently, that the principal driving force in energy policy is the environmental dimension.

Evolution of Energy Policy

As one looks back to the characteristics of energy policy—or perhaps more accurately, energy policies, as it is doubtful that the country has ever had a coherent, consistent policy—one can identify a number of components. Within the broad framework of "abundant, reliable, cheap" energy supplies as the objective of energy policy, emphasis changed significantly over time. An early strand was concern with the economics and, one must add, the policies of the domestic oil industry. Given the size of the major players and their potential for market manipulation and domination, it was not surprising that curbing the market power of oil was an important policy objective. However, in actual fact, major public policies were designed to aid rather than tame the industry. For instance, the depletion allowance gave the industry a tax advantage, the Connally or "hot oil" Act shored up the industry's ability to stabilize the price of oil, and so-called pro rationing schemes, established in the major producing states and complemented by estimates made by the Federal government of future supply "needs," all worked in the same direction, that is, to help the industry achieve a sustained level of profitability.

In the fifties, a new driver appeared in the form of the national security issue. Its aim was to slow down the rising share of imports in meeting domestic oil consumption. Import quotas were established to enable domestic producers to cope with competition from cheap imports, coming mostly from the rich, recently discovered and rapidly expanding production in the Middle East, and produced in significant degree by American companies. That policy structure, increasingly riddled by a host of exemptions and considerations of special situations, such as size, location and so on, gradually eroded. It collapsed for good just about the time that OPEC launched its offensive in the early seventies.

By that time, Congress had passed the comprehensive National Environmental Policy Act (NEPA) and established a new agency to implement its provisions—the Environmental Protection Agency (EPA). Additionally, it had created the Council on Environmental Quality (CEQ), with a mission that was, perhaps, too all-encompassing ever to be accomplished. Above all, it lacked clout. In any event, by the early seventies the link between energy and deterioration in the quality of the country's air, water and land resources was firmly established. The fact that all this was almost wholly missed in the book illustrates how rapidly events can and do modify the perception and direction of problems.

Going beyond oil, a review of the legislation on the books under successive presidents and Congresses in the seventies testifies to a sharp change in focus from matters of national security, industry performance and structure, and other earlier concerns (such as the "proper" uses of natural gas, the yardsticks for setting electric utility rates, the role of public power, the fairness of railroad tariffs vis-à-vis coal transportation and so on) towards environmental matters. In particular, legislation shifted towards energy conservation as no longer merely a way of reducing import dependence for oil and softening the impact on the country's balance of trade and payments, but as a way of reducing pressure on environmental resources.

One source of energy, however, went its own way, so to speak, in terms of policy—nuclear power. Here the environmental dimension was a strong component from the early days on. Yet, here too, the emphasis shifted somewhat under the impact of external events. Concern for safety, in several guises, was present from the start, though what was considered to be "safe" changed. Early on, for example, stress was laid on safeguarding aquatic life in the presence of hot water discharges from nuclear reactors. The welfare of the striped bass became a judicial celebrity; but as time went on, it turned out that the matter was manageable, and the striped bass as well as another fish, the snail darter that cost more than just headaches to the Tennessee Valley Authority in whose waters it lived, have had their day and have largely been forgotten.

Unpredictable events such as Three Mile Island and Chernobyl, endeavors to achieve a nuclear capacity by various countries whose political stability was in doubt, plus an unending stream of duly reported minor, though potentially disastrous accidents, have led to the fact that the major elements of nuclear power concern are today as they were two decades ago—reactor safety, proliferation, and nuclear waste management, with proliferation perhaps having receded somewhat in the relative ranking.

More striking is the change in emphasis when one looks at coal. Miners' health and safety and landscape destruction, mainly in the case of strip-mining, were the two classic major environmental ills when the world of coal first came under environmental scrutiny. Landscape destruction, while still a problem in some locations and circumstances, seems to have been remedied to the extent that it has greatly receded as a major issue. Much the same can be said of health and safety questions. This has not, however, produced an era of tranquillity for coal. As one concern receded, another emerged. Acid precipitation was the next major adverse concomitant of coal burning, only to be overtaken in the most recent past by carbon dioxide emissions and their global warming effect.

This is not the place to enlarge on any of these topics, on which there exists a vast body of literature. The necessarily brief review of evolving issues and concerns regarding energy, in which the environmental dimension is of relatively recent vintage, is mentioned here to (1) illustrate how rapidly problems can achieve prominence and then become more or less acceptable pieces of the landscape, and (2) express some astonishment, with hindsight, that by the end of the sixties, when *America's Changing Environment* was put together, the environmental aspects of energy use appear not to have triggered a specific treatment, however cursory.

Why was Energy Omitted?

One must ask, why this omission? For one, the energy scene in the late sixties was relatively tranquil. Prices were low, supply was ample, and the connection with the environment while evident on the demand/consumption side had not yet begun to seriously affect the sources of supply. Aside from damage from surface or strip mining of coal, energy production processes went on largely unimpeded and unquestioned. It is also possible that public policy was so completely absorbed by trying to keep energy producers, especially the oil barons, under scrutiny that environmental issues lacked a vocal constituency.

Failure to address energy as a unitary field is hardly redeemed by a short paragraph in a chapter called "Quantity and Quality of Resources" (pp. 107–130). And here, how skewed the outlook! Possibly driven by looking at a long trend of rapidly rising demand, and by a case of technological optimism, it foresees energy demand in the year 2000 as triple that prevailing at the time of writing, i.e., roughly the mid-sixties (p. 115). In fact, by 1990 U.S. energy demand was only about 50% above the level of 1965, and by the year 2000 it is unlikely to register much more than that. Hopes for liquefied and gasified coal, as for oil from shale and from bituminous sands have gone unfulfilled, and so has the idea that "we may well be entering an era of slowly declining energy costs" (p. 115). The eruption of the oil crisis prompted by OPEC in 1973, the unforeseen rapid increase in the cost of nuclear power due to continuous modifications, poor workmanship and many unexpected difficulties associated with the new technology which led, among other things, to long delays in construction, and the rising demand for energy in the developing countries all combined to dash the hope for cheap nuclear energy. (On the other hand, predictions that energy costs would go cataclysmically through the ceiling turned out to be equally misplaced.) Technological optimism led to regarding the possibility that a nuclear breeder might be developed by the mid-seventies as highly likely. Twenty-five years later it is further away than ever, underscoring the previously mentioned trap of optimistically dating successful commercial evolution of technologies that appear to merit swift application, on the basis of their scientific and engineering attractiveness.

We may well be in the same position today vis-à-vis fusion that we were three decades ago vis-à-vis the breeder. Was the remarkable dampening of demand for energy foreseeable in the sixties? One must doubt it, for its roots lie largely in the impact of rising energy costs, and history provided no paradigm for such a development. Indeed, one is reminded of predictions at the time that with the full blossoming of nuclear energy, including the breeder, electric power costs would become "too small to measure."

The Urban Future That Didn't Come

In contrast to the neglect of energy, *America's Changing Environment* devotes a good deal of attention to the "environment" in the more restricted literal sense, i.e., the immediate living space, rather than the large envelope of air, water and land in the aggregate.

Urbanization and life in the contemporary city play a large part in the volume. Two of the twenty papers deal specifically with "the city," not by characterizing and providing remedies for specific diagnosed ills, but rather in a more lofty way by discussing "New Towns" and "The Experimental City," including a suggestion that in the next three to four decades, i.e., by the year 2000, we might need fifty "new cities" (p. xxxv). The concomitant observation that "our existing cities should be redesigned and in large part rebuilt" has an equally utopian cast. From the vantage point of the early nineties one must judge that these ideas have almost totally failed to produce results. Some might even argue that we have regressed. A place like Reston, Virginia, conceived as a showcase of the "new town," has lost much if not most of that character, and has largely reverted to being just one more bedroom community in the Greater Washington metropolitan area. In a purely topographic sense, the issue can also be said to have narrowed in that today's attention is much less on the city as a whole than on the inner city and on coming to grips with its open sores: crime, drugs, homelessness, unemployment and the whole package of adversities connected with them.

A whole range of changing circumstances prompted the country to turn away from city planning. Above all, population pressures, partly the result of purely demographic factors (e.g., the post-World War II baby boom), partly arising from the rapidly growing productivity of American agriculture and the consequent migration from farm to city, and partly medical and sanitary progress resulting in longer life spans, left little time, resources, and inclination for innovation. Efforts to create "new cities" yielded to mostly shabby and ill-conceived public housing, as well as to greatly increased means of transportation, i.e., the spread of the automobile. Extending the city by way of suburbs was the preferred solution, necessitated also by shrinking household size, growth of one-person households, and the consequent need for more separate living space. Again, external, autonomous factors impinged upon the environment in scarcely predictable fashion, while the innovators' ideas remained on paper.

Were one to assemble a study of this kind today, in addressing the problems of the urban environment, one would, I suspect, place heavy emphasis on crime, drugs, smog, health, welfare and education, though one must observe at once that these are facets of life that would not have fallen within the "environment" area in the 1960s, and may not, even today, be properly designated environmental concerns. Yet these are the elements we have in mind when we speak of the "inner city environment." A look at the volume's index finds no mention of either crime or drugs, while the two citations under "smog" (p. 278) refer to passages in a chapter on education. The city is dealt with more in an architectural, organizational, town-planning mindset than as a complex of physical, social and economic factors, some adverse, some advantageous, some neutral.

With a bit of exaggeration, one might be tempted to say that public concern has moved from a nearly "romantic" perception to a cold-blooded chemophysical one. The Clean Air Act and its amendments are silent on urban values of the kind that fascinate the authors of the relevant chapters. Today's regulations talk of permissible levels of emissions of various noxious gases, of entire metropolitan areas, of technological remedies and fixes (e.g., a minimum of 2 percent of all new cars sold in California in 1998 must be of the nonpolluting kind). They are unrelated to the nature of a specific city site, silent on historic buildings, neighborhoods, etc. They are a far cry from novel transportation schemes, such as those Athelstan Spilhaus, the geophysicist and fearless advocate, suggests in his chapter on the "Experimental City" (pp. 219–231)—mass transportation via small, electricity-driven and computer-controlled "pods" with propulsion in the track, that would be small enough to "pass noiselessly through buildings with normal ceiling heights" (p. 227). Connections to intercity travel would be provided at the city's periphery, and only there. Needless to stress, such a system would function only in a new city built specifically to accommodate it. Unfortunately, the reality today is that we must be content with reducing the poisonous fraction of the city air we breathe.

While we have given up on novel transportation schemes for the city, science fiction has become reality in space travel. Man on the moon is more exciting than man in the city. Obviously, engineering missions, no matter how complex, are more easily accomplished (and far more rewarding to the sponsors and participants) than changes that re-engineer the societal micro-cosm. People and society are vastly more complex and less easily reduced to numbers than physical phenomena so that we understand the effects of human intervention on physical phenomena much better than the effects of social engineering. Hence, the futility of the conventional lament: if we can put a man on the moon, why can't we abolish poverty, and so on.

The Judiciary and its Alternatives

Another major theme missing in the collection of essays is the role of the Judiciary and, more broadly, the search for institutional "handles." None of the chapters foreshadow the veritable explosion of litigation that has become a hallmark of environmental development. Why this gap? Most likely because the magna carta of environmental affairs, the National Environmental Policy Act, did not come into force until 1970, and because the tidal wave of litigation aimed at establishing the "proper" interpretation of the Act's components (most prominently, when is an environmental impact statement required and what should it contain?) did not strike until the early seventies. Nor has it ceased since, to the extent that it is now almost an article of faith that any new statute passed by the Congress on environmental affairs, or the more important rules and regulations issued by the EPA and other relevant regulatory bodies, will be subjected to review by the courts, or, at a minimum, will be greeted by the threat of litigation as a means of inducing the promulgating agency into modifying its stance.

In lieu of judicial action, various authors recommend new institutions to assure attention to such aspects as cannot be satisfactorily translated into dollars and cents—the "intangibles." Nathaniel Wolman, for example, proposes that all agencies concerned with the use of resources, public construction and zoning create the staff position of "aesthetician" (p. 144).

In a similar vein, he proposes that a "network of citizen groups be created…to serve as a link between the experts and the electorate: urban, rural, county, state, regional, federal, landscape, water, air, wilderness, architectural, mountain and so forth" (p. 143). What has occurred in the intervening quarter of a century, instead of such a profusion of overlapping "networks," is the flowering of what have come to be called special interest or single purpose groups, too numerous to name here. These were not, of course, created by an act of policy-making. Noteworthy are a handful of major environmental advocate organizations (the EDF, NRDC, Sierra Club, etc.) that, for the most part, operate across the board and have become the primary initiators of litigation. Most recently, they have begun to add to their legal arsenal the attempt to form alliances with the generators of pollution or other environmentally adverse phenomena as a means of achieving mutually acceptable objectives that would make litigation unnecessary.

This whole field, so important in decision-making today, is a major lacuna in the volume. This in spite of the right questions being asked, such as in Wolman's chapter:

> The new economics poses for itself essentially political problems. Which decision shall be made through the market and which not? What decision structure will best assure environmental quality? How can aesthetic feelings be translated into public policy? What happens if strong aesthetic impulses are shared by only a small minority? Consideration of the problem surrounding public preferences will help throw the many political dilemmas into sharper focus (pp. 144–145).

These queries which could have been written today, rightly envision a vast research field, the findings of which have not by a long shot exhausted the opportunities and possibilities for useful answers. But even in such insightful writing the role of the courts is curiously omitted. Indirectly, Roger Revelle, dealing with recreation facilities, hints at it in his own graphic language:

> Recreation advocates must be able to convince a majority of their fellow citizens that the value of a recreational activity will be greater than the values it replaces, or else *make such a nuisance of themselves* that the line of least resistance for the majority is to placate them…. Here as in many other human affairs, passions spin the plot (p. 269) (emphasis added).

Micro- and Macro-Environment

A generalization occurs at this point. Would it be useful to distinguish between micro- and macro-environment, where the former is defined as the immediate surroundings, the habitat that shapes and conditions the individual's comfort and welfare, the house a person lives in, the street on which it stands, the noise level, the maintenance of it all, which is where, for the most part, America's Changing Environment lives, while the latter deals with the "biggies": air pollution and water pollution on a large scale, for example. The point is arguable. What matters and gives rise to concern and policy implications is surely the immediate surroundings, which include foul air as much as noise. What has happened in the past two decades is that aspects of the environment have emerged that were neither obvious nor detectable by the individual, but are in the long run perhaps more insidious. A case in point: Azriel Teller's chapter on air pollution argues against national quality and emission standards and pleads for treatment of small units. "There can be neither a national blueprint for solving the air pollution problem nor national emission standards. Each city, or air-shed must approach air pollution with respect to its own particular situation...." (p. 54). Thus acid rain, for example, an environmental topic of regional scope writ large in the past two decades, and not surprisingly, global warming, the latest entrant on the environmental stage, receive no mention.

Assuming that the volume is representative of thinking on environmental problems at the time in which it was put together, one might venture to propose that we have moved from emphasizing the way the individual or the small community is affected to looking much more impersonally and abstractly at large systems, consideration of which, to be sure, is triggered by their impact on individuals, but moves on from there to lead a life apart in the form of large systems, national, regional and global.

Does this suggest that we have come, deliberately or inadvertently, to ignore the microdimensions? Yes and no. Yes because it turns out that they are usually local manifestations of phenomena with wider bearing and lend themselves to more efficient and effective treatment on the larger scale. Yes, also, because analysts and policy-makers find that they often know more about the workings of larger systems than those on the individual or micro-scale, or at least, that larger systems are more easily dissected and structured for research. No because what matters in the end is the specific impact of the larger system, and of its changes, on the smallest entity, be that a human being, an endangered species or a given tract of natural beauty.

Changing Research Approaches

Altogether then, the focus of interest of scholarly research seems to have shifted from contemplating the microscene to addressing large systems. This has been accompanied, though not necessarily as a consequence, by a change in tools of analysis. I am thinking here above all of the pervasiveness of mathematical modeling. Leaf through the pages of the book, and you will find no equations, no integral symbols, no higher mathematics. A far cry from the bulk of contemporary treatment of environmental issues. The volume's approach is largely an "aesthetic one," in the best sense of the term. An example of this is the following paragraph in the Introduction, written, I remember well, by Roger Revelle:

Changes in population density and distribution have been even more destructive of environmental values. For one who lived in and loved California in the 1920's and 1930's, the fourfold growth of the state's population since 1930 can be thought of only as a disaster. Instead of the brilliantly clean air, vivid blue skies, and sharp-edged mountains of those days, there is now a brownish, opaque, and acrid veil of smog over vast areas. Gently rolling wheat fields and orange groves have been replaced by an endless monotony of tract houses, shopping centers, and "roadside businesses" of such an ugliness that they dull the eyes and cannot be perceived in detail, but only as a blur. Through the wonders of modern earth-moving machinery, the soft brown and green oak-covered hills have been "recontoured" into flat building sites and badlands slopes, and cut and filled for the straight scars of freeways more deeply than one would have believed possible (p. xix).

The volume abounds in such observations and language. A nostalgic aura of paradise lost, or about to be lost, pervades much of it. We are at risk not so much because we breathe foul air, but because we sink into ugliness. While not amenable to true measurement, the aesthetic approach conveys orders of magnitude and intensity that speak to the nontechnical reader—a significant merit when one writes about the environment. Moreover, we are often fooled into thinking we have taken the measure of a problem once we have given it mathematical clothing, forgetting that lack of knowledge and/or data renders the formal approach incomplete and shrouded in uncertainty. (The global warming debate owes much of its fascination precisely to the phenomenon that the entire chain from data to understanding the impact is beset by uncertainties.)

Changing Agenda and Focus

As with everything, there are fashions. Agenda and priorities change over time. Roughly speaking, outdoor recreation, new urban forms and landscape planning, prominent in the volume, seem to have moved down on the priority list. As mentioned above, unemployment, crime, drugs and so on have moved up and now constitute the core of the "urban environment problem."

Conventional, traditional environmental topics—land, air and water—continue to occupy an important place on the "worry agenda," though with ups and downs dictated largely by external events. That is to say, we tend to react easily to the emergence of palpable problems. Arguably, environmental concern as a public matter began with foam from detergents floating in the nation's streams and with Rachel Carson's *Silent Spring* (1963). More generally, prolonged periods of drought and massive spills or emissions of unwanted substances, are apt to push water pollution and scarcity problems to the forefront of environmental concern. Offshore activities, such as drilling for oil constitute a perennial concern, with emphasis changing according to levels of activity on a geographic scale (e.g., California's coastline, Alaska's North Shore, etc.). Incidentally, somewhat surprisingly, the volume fails to look at offshore drilling, perhaps because it had not, at the time, become an environmental battleground. The major event that triggered attention—the Santa Barbara oil spill—did not occur until 1969. This is in sharp contrast to the extensive treatment of the movement then flourishing to save the San Francisco Bay from land-filling.

There are other notable blanks. Most striking, as mentioned above, energy as a separate topic, and within it, oil, gas, coal and nuclear power, the last of which curiously appears only as a factor in the preservation of highly-valued landscapes, the case in point being the then pending proposal by Pacific Gas and Electric to build a nuclear power plant at Bodega Head or Diablo Canyon (pp. 69–71).

Technological optimism is at work again in projecting the demand/supply balance outlook for metals. Unconventional sources such as the ocean floor are suggested as means of preventing or easing any supply constraints in the foreseeable future. They have, of course not materialized to this day, though the general judgment that supply constraints would not impede economic growth has proved correct, and such perceptions, rampant in the 1970s and 1980s, as the "resource war" in which the adversaries of the United States would cut it off from vital supplies of "critical and strategic materials" have proved unproductive fantasies. "In our lifetime and those of our children the availability of energy and raw materials will not put any serious limits on our prosperity" reads a sentence in the Introduction. That has proved to be a durable piece of wisdom. But what has made for supply sufficiency has been steady advances in conventional technology that have brought into production vast new resources of lower and lower grade, rather than quantum leaps into new worlds, be these on the ocean floor or on the surface of the moon.

In Search of Major Themes

What can one distill as some of the major themes of this collection of essays? In trying to answer that query, we discover that these themes are still right on target and that we seem to have made some progress moving towards better understanding and, perhaps, management of the environment. One is our ignorance of physical interactions that take place in the environment, then highlighted in the field of air pollution, and today, on an even greater scale, in the global warming phenomenon. Decision-making in the presence of inadequate knowledge is another theme. Recalling that the papers were written several years before Congress passed the National Environmental Policy Act (1969) and the establishment of the EPA (1970), that theme was even more relevant, as there was no focal point where the search for knowledge could be triggered and the results collected, analyzed and channeled toward action.

Deficiency of knowledge of the workings of the physical universe is constantly being remedied, but as for knowledge of what society wants collectively and how much it is prepared to pay for it, progress has been slow; perhaps more rapid in the formation of new institutions than in providing the knowledge needed by those institutions to render decisions. "Invention of tools for economic measurement of human sensory or emotional delight and deprivation is at an early stage" (p. xxiii), the Introduction rightfully observes (and could observe if the piece were written today), but how the courts assumed a major role and attempted to fill the gap was not foreshadowed in the mid-sixties. Instead, more than one of the authors postulated that while economics may help illuminate the merits of multiple competing claims, decisions are ultimately made in the political arena.

Summing Up

What strikes me above all on rereading *America's Changing Environment* is the transformation of the "environmental problem" from one concerned primarily with aesthetics, with near utopian visions, and with neighborhood matters to issues of public health, deterioration of the basic life support systems, and broad international and global aspects. Indeed, the volume is engaging because of a certain intimacy that is apt to catch the reader's attention and sympathy. To wit, when Roger Revelle writes about better use of city parks, he offers a whole raft of potential activities, ranging from cultural events to athletics, to demonstration farms and mills, to winter fishing, and last but not least, to "love-making, for which purpose benches and movable chairs should be provided, as well as privacy and protection against criminals...." (p. 259) Vintage Revelle, but also symptomatic of the focus on detail, small scale, individual well-being. Contrast this with today's stress on large-scale systems and their physical and chemical characteristics.

One factor in this transition is no doubt the recognition that many major problems do not respect borders and that local remedies tend to be futile. Moreover, some environmental problems have long gestation periods and thus do not lend themselves to short term analysis and policies.

In this context the Clean Air Act of 1970 marks the transition from dealing with micro- to tackling macro-environmental problems. These trends have now coalesced in the concept of "sustainable development," presumably applicable to the whole range of environmental issues and finding their most recent guise in the problem of global warming. It is indeed a long way from neighborhood parks to near cosmic phenomena, and one must wonder what new transformations lie ahead of which we are now as unaware as were the authors of this volume of many of the problems of the 'nineties. It was the genius of the man this volume honors that his hunger for knowledge enabled him to span the entire range of problems and infuse it with enthusiasm.

References

Carlson, R. *Silent Spring*, Boston: Houghton Mifflin Press, 1963.

Landsberg, H. "Energy in Transition, The View From 1960." Energy Journal, vol. 5, no. 2 (1985).

Landsberg, H, L. Fischman and J. Fisher. *Resources in America's Future*. Baltimore: The Johns Hopkins Press, 1963.

Resources for Freedom, 35th Anniversary edition, with a foreword by William S. Paley and an afterword by Hans H. Landsberg, "Resources for the Future 1952–1987." Washington, DC, June 1987.

Revelle, R. and H. Landsberg, *America's Changing Environment*. Beacon Press, 1970.

Notes

1. This paper has greatly benefited from the suggestions of Robert Dorfman.

5

Population and Environmental Deterioration: A Comparison of Conventional Models and a New Paradigm

By Peter P. Rogers[1]*

Introduction

In a prescient paper prepared over 20 years ago for the 1974 World Population Conference in Bucharest, Roger Revelle (1974b), aided by Harold Thomas, identified three sets of environmental problems with important demographic interactions. The first set of problems are those that spring from poverty and inadequate social and economic development. These problems are mostly experienced in the developing countries. They typically result from widespread unemployment and underemployment in the rural areas which force the poor into the cities, from the need for increasing agricultural production to meet the needs of rapidly growing populations, and from the lack of financial resources and institutions to build and manage the environmental infrastructure required to mitigate harmful consequences. Rapid population growth makes all of these effects temporally more pressing. Revelle characterized this first set of environmental problems as the "pollution of poverty." At the Stockholm conference in 1972, Indira Gandhi said that "Poverty is the biggest polluter."

The second kind of environmental problems is felt mainly by the industrialized countries and in the handful of rapidly industrializing developing countries. In these cases the problems arise from the high levels of industrial production which require ever increasing amounts of energy and other raw materials. The industrialized countries' situation is exacerbated by the high levels of personal consumption of goods and services by the entire population. The effects of these problems on health and well-being are quite different from, and typically much smaller than, those of the first kind of problems. Revelle characterized this second set of problems as the "pollution of affluence." Absolute population size generates these problems directly in proportion to size, but the degree of affluence of the population is what makes the large difference from problems of the first kind.

The third kind of problem deals with global environmental change brought about by the cumulative effects of human activities. These types of effects, which dominate current debate about the environment, were unheard of 20 years ago. Revelle, through his work on carbon dioxide in the oceans and the atmosphere, was among the first scientists to sound the alarm about the potential for large-scale, and possibly unpleasant, change in the global environment as a result of small, barely perceptible increments to global cycling of various benign as well as noxious chemical compounds. Here both population size and affluence play important roles. The first two kinds of problems are essentially "flow" problems, where the impacts are largely due to the flow of materials and residuals through the ambient environment. The third kind of problem is associated with the "stock" of materials and residuals building up in the environment. Hence, the resolution of this type of problem will require more than merely a reduction in the flows; it will also require some reduction in the current or future stocks. These are inherently more difficult problems to solve.

This paper reviews the original formulations and finds that they have held up remarkably well during the intervening 20 years. In addition the paper reviews recent evidence about the population, economic growth and environmental deterioration and concludes that the role of economic growth in mediating environmental deterioration may have been seriously underestimated.

The Predictions

Revelle was not particularly sanguine about the prospects for rapid improvement of the first kind of environmental problem. He focused his attention on water and sanitation and urbanization problems in this category, concluding that: "If waste disposal problems of the rapidly growing, densely crowded, poverty-stricken cities of the developing countries of Asia, Africa, and Latin America are to be solved, new and ingenious engineering solutions are called for." (Revelle 1974, 17)

For the second kind of environmental problems, Revelle predicted the rapid improvement of the environment that we have seen in the wealthy industrialized countries:

> If this rate of increase in the level of abatement continues and can be extended to cover the entire range of the processes of environmental deterioration, such deterioration in the developed countries may prove to be a transitory phenomenon. Our descendants may wonder at the intensity of our present environmental concerns! (Revelle 1974, 17)

Twenty years ago Revelle had little to offer with regard to the amelioration of problems of the third type: global environmental problems. He did say that human populations might be able to adjust to the changed conditions less painfully if realistic forecasts could be made. He suggested that in order to do this "international cooperation in monitoring and analysis is essential." Since most causes of global change are due to exploitation of energy sources of one kind or another, he urged extensive research and development to develop solar energy sources.

> Most potential climate effects could be avoided if sunlight could be utilized as the principal source of energy for human use. Research and development of solar energy should be given high priority in mankind's continuing search for new energy sources (Revelle 1974, 22).

The lines of action he proposed to reduce the environmental deterioration due to human energy use were:

1. reduction of total energy use;

2. introduction of nonpolluting energy converters;

3. removal of noxious substances before they enter the environment; and

4. substitution of non-polluting energy sources for those that pollute (Revelle 1974, 10).

What Actually Happened

As Revelle predicted, the environmental conditions in the developing countries have continued to deteriorate and those in the developed countries (leaving aside the former communist countries) have continued to improve. For example, despite the 1980s having been the UN's International Drinking Water Supply and Sanitation Decade, large numbers of the world's population are still without adequate water and sanitation and the effort to reach these people is hindered by the continued rapid growth of population and the shortage of capital investment funds.

The UN and its specialized agencies played a major role in the International Drinking Water Supply and Sanitation Decade (1981–1990), which had the overly ambitious goal of providing clean water and adequate sanitation to all by the year 1990. Christmas and de Rooy (1991) provide a summary of the accomplishments of the Decade. Whereas in 1980 only 40% of the world's population had access to a safe water supply and only 25% had access to sanitation (2 billion people lacking both), by 1990 over 60% had access to a safe water supply and 40% to sanitation. This represents a major achievement especially given the growth in population. Surprisingly, more success was achieved in rural than in urban areas. According to Christmas and de Rooy (1991), by 1990, 82% of the urban populations had access to potable water and 72% to some form of sanitation. Again, these global figures cover a wide range of country experiences.

The New Delhi Statement (Water International 1991) compares the current $10 billion annual capital expenditures with the estimated $50 billion annual expenditures needed to achieve 100% coverage for both water and sanitation by the year 2000. Munasinghe (1990) points out that bilateral and multilateral lending in this area has traditionally been less than $1 billion per year. He concludes that, since much higher levels of finance are unlikely to be forthcoming, the shortfall will have to be met by greatly improved cost recovery from new and existing projects.

Neither Revelle nor anyone else predicted the high levels of environmental degradation that were created in the centrally planned industrial countries. Revelle, however, would no doubt have agreed that even in these countries, the deterioration is likely to be a transitory problem. Nor did Revelle predict the intensity of attention that the third set of issues has received in all parts of the world. During June 1992 the UN-sponsored Earth

Summit (UNCED) was held in Brazil, attended by the heads of state of more than 100 countries. The focus of this meeting was almost exclusively on global issues. Unfortunately, the population dimension received scant attention. One of the centerpieces was the issue of global warming, with an attempt to negotiate a treaty that would bind governments to specific emission policies.

Despite the emphasis on problems of the third kind at the global summit, the World Bank in its 1992 Development Report, entitled *Development and the Environment*, essentially reinvented Revelle's 1974 set of priorities when it said that "setting environmental priorities inevitably involves choices. Developing countries should give priority to addressing the risks to health and economic productivity associated with dirty water, inadequate sanitation, air pollution and land degradation which cause illness and death on an enormous scale" (World Bank 1992, 44).

Global climate change concerns ranked last in the Bank's list of priorities, and its action recommendations in that area, almost identical to Revelle's, were limited to information and research coupled with the development of renewable energy sources. This contrast between UNCED and the World Bank is remarkable inasmuch as it shows which best reflected the real needs of developing countries. It appears that UNCED was dominated by environmental interest groups from the industrialized countries, while the developing countries blocked discussion of population as a leading cause of environmental deterioration. The result of this maneuvering and posturing was a high level of frustration experienced by many participants.

The Change from Pollution Clean-Up to Ecosystem Maintenance

There seems to have been a major revision of what is meant by the term "environment." In the early 1970s environment essentially meant local air, water, land and soils which were characterized as "polluted" or not. By the start of the 1990s environment meant "ecosystem." This is a major conceptual leap, moving from dealing with the impacts of humankind's effluvia upon itself to the impact upon the entire ecosystem. This revision of the definition of environment has led to many difficulties in communication about the environment. There is a major difference between making the environment safe for present generation of humans and making the ecosystem "sustainable" for the future. The Brundtland Commission enshrined the concept of "sustainable development" in the lexicon of the United Nations' bureaucracy: "Sustainable development is development that meets the needs of the present without compromising the ability of future generations to meet their own needs" (World Commission on Environment and Development 1987, 43).

The concept of sustainable development has been taken up by many environmentalists and others who insist that development must be sustainable with respect to ecosystem maintenance. This will be a difficult task unless there are major efforts to control and reduce human populations, as is demonstrated by Table 1. Table 1 shows the UN's best estimates (Population Reference Bureau 1991) of the amount of time required to add an additional 1 billion people to the earth's population. It took all of human history until 1800 to reach the first 1 billion people living at one time on the globe. To reach the next billion took 130 years. To reach the third billion took only 30 years, the fourth 15 years and the fifth only 12 years. When this pattern of growth is projected into the future, we see that it will take only about an additional 11 years to reach each of the sixth, seventh and eighth billions. After 2020, the projected intervals lengthen again, until it takes 34 years to reach the twelfth billion in 2110. This interval is similar to the interval between the second and the third billion. Given these projections it is difficult to conceive of any sort of dynamic equilibrium between humankind and nature during the coming six decades.

Even if the population numbers themselves could be stabilized, there is no guarantee that the demands placed upon resource use and the environment

T A B L E   I

World Population: Number of Years to Add Each Billion

Year	Years to Add
First billion reached 1800	all of human history
Second 1930	130
Third 1960	30
Fourth 1975	15
Fifth 1987	12
projected	
Sixth 1998	11
Seventh 2009	11
Eighth 2020	11
Ninth 2033	13
Tenth 2046	13
Eleventh 2066	20
Twelfth about 2110	34

Source: Population Reference Bureau (1991), based on United Nations and World Bank estimates and projections.

would also be stabilized. In order for that to happen economic growth in the developing world would have to halt. The reason economic growth would have to halt only in the developing world is discussed at the end of the next section.

The shift of the concerns of environmental policy over the past 20 years from reduction of anthropogenic pollution to ecosystem maintenance has not only raised the goalposts, it has also made them highly moveable. Cleaning up human pollution is a difficult and costly task, but it can be understood, planned for and implemented over a reasonable time horizon. Ecosystem maintenance, or the more commonly used term "sustainability," is a much more diffuse and poorly understood concept. Ruttan's (1991) discussion of sustainable growth in agriculture is one of the more helpful analyses of the concept of sustainability in the literature. Starting with a review of historical examples of sustainable agricultural development, Ruttan concludes that none of the traditional systems is capable of responding to modern rates of demand growth. He argues that because of the tremendous achievements of agricultural research and the technical and institutional innovations in many countries, food supplies can be considered sustainable for the foreseeable future as long as investments are targeted toward maintaining the agricultural research system worldwide. In reaching this conclusion, Ruttan shifts the meaning of sustainability from the ecosystem itself and focuses it on the continued ability to meet human needs for food.

Models of Population-Environment Interaction

In his paper, Revelle (1974a) attempted to explain the interrelationships among population growth, economic growth, and environmental deterioration. He was bothered by the simplistic identity, developed in the early 1970s by Commoner and Ehrlich, in which the environmental disruption in a given region was derived from the product of the size of the population, the level of consumption per person and the technology used. This relationship has been variously formulated, but one version is that, for a given region, the wastes produced are identical to

$$w = P \, (GRP/P) \, (w/GRP) \tag{1}$$

where
    w = quantity of wastes produced in region;
    P = population of the area; and
    GRP = gross regional product.

In other words, waste generation can be divided into three components: population, per capita income and the technology of waste production.

Revelle emphasized that the ratios on the right hand side of equation (1) are not constant but are determined by a set of other relationships. These concerns led Revelle to consider a more complex formulation, which allowed for explicit substitution between production and treatment technology by introducing explicit functional relationships as follows:

■ the quantity of wastes produced is a function of the production and consumption technologies in use and the size of the gross regional product

$$w = a(GRP)^B \, T \qquad (2)$$

■ the ambient pollution level is a function of the quantity of wastes produced, the level of abatement applied to these wastes and the area into which the wastes are emitted after treatment

$$W = b[w(1-r)/A]^{B1} \qquad (3)$$

■ the total impact upon human society is a function of the ambient levels and the number of persons exposed to those levels

$$I = cPW \qquad (4)$$

■ and the waste treatment levels chosen are a function of the gross regional production per capita, the amounts of wastes produced, and calendar time

$$r = f(GRP/P, \, t, \, w) \qquad (5)$$

where:
  w = quantity of wastes produced by industry and society in region;
  W = ambient pollution levels caused by wastes entering the environment;
  T = measure of the technology used;
  A = area into which the wastes are disposed;
  P = population of the area;
  I = total impact of environmental pollution on human beings;
  t = time;
  r = treatment level;
  GRP = gross regional product;

b = a factor reflecting diminishing returns which increase waste production from natural resource use as GRP increases;

b_1 = an absorptive capacity factor which allows the environment to absorb a certain level of wastes without much resulting pollution;

a, b, c are constants, the values of which can be varied depending upon human behavior.

Revelle was unable to estimate his model from data then available. Twenty years later we are still not able to obtain empirical estimates of the parameters in his model because cross-sectional or longitudinal data are not recorded for many of the variables that he suggested. It may surprise many people, as it certainly surprised me, that despite the large amount of resources devoted to gathering data on environmental conditions in many countries around the world (see WRI 1986, 1987, 1990, 1992; World Bank 1992; Asian Development Bank 1990, 1991; ESCAP 1992; UNEP 1987, 1989; and many country-specific reports from developed and less-developed countries) no reliable time series of environmental quality data consistent with economic and population data seem to be available outside the OECD countries. Even for those countries the time series do not go back far enough to capture some of the interesting developments discussed below, or they are not easily comparable among the countries.

Moreover, it is not clear that there is any particular nobility to the model specified by Revelle. It does mix up equations describing the structure of the economy-environment interactions with other equations describing the dynamics of the flow of materials through the economic system. Moreover, since T is not a scaler quantity, f is not a well-defined function, and w and W cannot be readily measured for aggregates of pollutants, there is little reason to think that this model will work for aggregations of pollutants, or for pollutants considered separately.

As an example of a different type of model specification which deals with disaggregated pollutants, Ridker (1972) completed a 185 sector dynamic input/output model of the U.S. economy in 1972 which predicted resource use and pollution from 1970 until 2000. Ridker's approach is similar to that put forward first by Leontief (1970) and later by Isard (1972), and used more recently by Repetto et al. (1989) and Jorgensen and Wilcoxen (1991), among many others. Ridker used his model to estimate resource and pollution elasticities with respect to population and GNP per capita by running the model separately with a one percent reduction in population and a one percent reduction in per capita GNP, both at the year 2000. Table 2 shows the resource elasticities for various minerals expected to be used in the U.S. economy in the year 2000. In each case, the population elasticity is less than

Resource Elasticities with Respect to Population and GNP per Capita in 2000[a]

Resources	Elasticities	
	% Population	% GNP per Capita
Iron	0.56	0.85
Aluminum	0.48	0.88
Copper	0.56	0.97
Lead	0.46	0.79
Zinc	0.64	0.99
Manganese	0.58	0.82
Chromium	0.51	0.87
Nickel	0.51	0.87
Tungsten	0.55	0.81
Molybdenum	0.61	0.88
Vanadium	0.61	0.83
Cobalt	0.64	1.40
Tin	0.22	0.68
Magnesium	0.56	2.90
Titanium	0.46	3.59
Sulfur	0.56	1.76
Phosphorus	0.74	0.92
Potassium	0.46	0.92
Nitrogen	0.46	0.93

[a] Percent change in resource requirements associated with a one percent change in population or a one percent change in GNP per capita.

Source: Ridker (1972, 45)

two-thirds the GNP elasticity. Assuming that the GNP per capita grows at least as fast as the population, this implies that the preponderant effect on resource consumption in the U.S. will be economic growth rather than population growth. Similarly, Table 3 shows that for all but one pollutant (carbon monoxide) the population elasticities are significantly lower than the GNP per capita elasticities, also implying that, for the U.S., income growth will have a much larger impact on pollution than population growth.

However, Ridker's and the other models that use Input/Output analysis require strong assumptions regarding the structure and development of the economy, and immense amounts of information. It is unlikely that such information will be available for many economies any time in the near future. Moreover, the model's detail may inhibit its use for the purposes that Revelle

Pollution Elasticities with Respect to Population and GNP per Capita in 2000[a]

| | Elasticity | |
Pollutant	% Population	% GNP per Capita
Particulates	0.46	0.78
Hydrocarbons	0.29	0.59
Sulfur oxides	0.42	0.87
Carbon monoxide	0.26	0.04
Nitrogen oxides	0.29	0.31
Biological oxygen demand	0.55	0.67
Suspended solids	0.53	0.71
Dissolved solids	0.56	0.69

[a] Percentage change in a particular item associated with a one percent change in the column heading.

Source: Ridker (1972, 47)

had in mind. For macro-level models, Slade (1987, 364) provides an excellent review of the natural resources/population nexus and draws some strong and fairly negative conclusions:

> It should be abundantly clear that all attempts here to draw conclusions about the effects of resource scarcity on the growth of population and economic well-being at the macro level failed. Keyfitz (1991), on the other hand, believes that simple macro models do have validity if only in pointing out the magnitudes of the problems (even in cases where the models themselves are incorrectly specified).

Revelle suggested simplifying his model to something akin to equation (2) above. The example he gave was that of urban sewage. The model is:

$$BOD = a\,P\,(GRP/P)^{\beta} \qquad (6)$$

or

$$BOD = a\,P^{1-\beta}\,(GRP)^{\beta} \qquad (6a)$$

where:

BOD = Biochemical Oxygen Demand, the quantity of oxygen required to oxidize the sewage waste in kg/day.

Revelle and Thomas applied this model to the city of Seoul, South Korea, where the parameter "a" was estimated as 0.0162, and β was 0.374. This implies a BOD elasticity of 0.626 with respect to population and 0.374 with respect to GNP per capita—the opposite of the Ridker results for the U.S., where BOD had a higher elasticity with GNP per capita than with population. One is tempted to believe that this switching of the values of the elasticities might be a generalizable difference between the developed and the less developed countries; however, one single observation is not sufficient to test a theory. Further, as discussed below, Ridker's results do not hold for the current U.S. situation.

A New Paradigm

Recent thinking about the population-environment nexus suggests that there is an inverse U-shaped relationship between per capita income and environmental insult. Bernstam (1990) articulated this relationship for energy resources, and the World Bank in its *World Development Report* (1992) claims that these relationships hold for many conventional pollutants. Figures 1 and 2, taken from that report and based upon empirical estimates by Shafik and Bandyopadhyay (1992), show how the new paradigm works. Figure 1 shows emission data schematically over calendar time for the industrialized (OECD) countries, and Figure 2 shows the agreement of the paradigm tested with data on sanitation coverage, SO_x concentrations, and per capita income for a range of urban areas in different countries (the data are from the World Bank data base and include between 55 and 70 countries for country data and 35 cities from 21 countries for the urban data. These are not time series but cross-sections for the years indicated. See Shafik and Bandyopadhyay (1992) for the details). In the top panel of the figure cross-sectional data for sanitation coverage for cities for two time periods are reported. The data indicate a strong relationship between per capita income and urban sanitation coverage. Moreover, the data for the later time period seem to indicate that there has been an improvement for the same levels of per capita income at the later date. The lower panel, however, is a more striking example of the new paradigm. At low levels of income, the concentrations of sulfur dioxide in urban air increase, reach a maximum, and then decline rapidly as income per capita increases. These data also seem to reflect a temporal downward shift between 1976 and 1985.

Based on these two figures, one can posit at least four mechanisms at work. First, some sort of technological change is taking place over time, which leads to the adoption of more efficient resource use and, hence, to less emission of residuals into the environment, at each level of per capita income. Second, as

FIGURE I

Breaking the Link Between Growth in GDP and Pollution

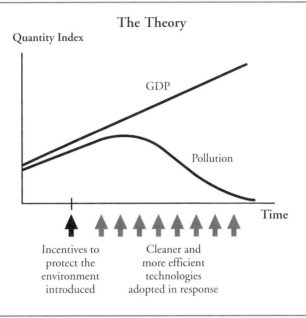

The Theory

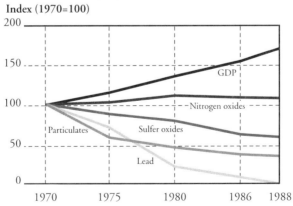

The Practice: GDP and emissions in OECD countries

Note: GDP, emissions of nitrogen oxides, and emissions of sulfur oxides are OECD averages. Emissions of particulates are estimated from the average for Germany, Italy, Netherlands, United Kingdom, and Unites States. Lead emissions are for United States.

Source: OCED 1991; U.S. Environmental Protection Agency 1991.

incomes rise over time, demand for environmental quality increases, leading to governmental regulation to encourage quality improvement. Third, at the same time, the scarcity of resource inputs becomes more accurately reflected in their prices leading to innovative behavior and a reduction of residuals emitted. Fourth, due to rising incomes the economic resources available for investment also increase, making government action less onerous and also providing the capital resources to underwrite the efficiency improvements.

We see that there must be some sort of technical change from Figure 2. In both examples, the environment improved for a given level of per capita income between 1976 and 1985. Thus, even someone living in a country with a per capita income level of $500 is likely to breathe cleaner air than people in previous decades in countries with the same per capita income level. Technology is copied from one country to the next, efficiency improvements are often incorporated without any ostensible incentive to do so, imported technology tends to have a higher level of technical sophistication than indigenously produced goods. Of course, this technical change may also be forced by increased scarcity on the input side which could lead to incentives to economize without any increased willingness to pay on the part of the consumers.

The lower panel in Figure 1 shows that incentives or regulation must also have their impacts. The incentives on the input side of the economy seem to be more effective than those on the output side. In other words, the effect of increased energy pricing in the OECD countries has led to dramatic declines in the energy use per dollar of economic output. In turn this has led to major decreases in the effluents normally associated with energy consumption, air pollution and major reductions in waste-water production because of improvements in capturing the heat in process water. As Shafik and Bandyopadhyay (1992) point out, almost all pollutants seem to be affected by this relationship. They claim that exceptions are biochemical oxygen demand in rivers, municipal wastes, and carbon dioxide emissions. (It is not clear why the carbon dioxide would behave this way since reduction in fossil fuel consumption will lead directly to reductions in the carbon dioxide emissions.) They also computed environmental elasticities with respect to income for three of their environmental measures over an income range from $500 to $11,000 per capita. For suspended particulate matter in the air, the elasticities ranged from a high of 1.4 to a low of -1.2, for sulfur the range was from 2.0 to -1.5, and for fecal coliform in the water it decreased from 7.0 to -3.5 and went back up again to 5.0. These elasticities are much larger in absolute values than those computed by Ridker (see Table 3) implying much larger responses to economic growth.

FIGURE 2

Changes in Urban Sanitation and Sulfur Dioxide Concentrations over Time at Different Country Income Levels

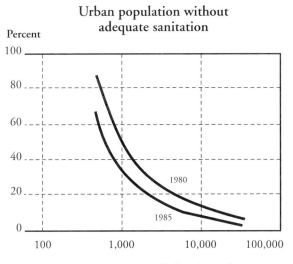

Urban population without adequate sanitation

Percent

Income per capita (dollars, log scale)

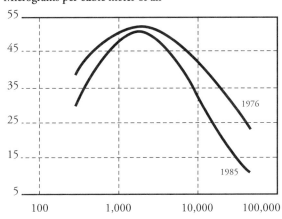

Concentrations of sulfur dioxide

Micrograms per cubic meter of air

Income per capita (dollars, log scale)

Source: Shafik and Bandyopadhyay, 1992 background paper, World Bank data.

The effect of governmental regulation on the environment is not clear from the data presented by the World Bank. Insofar as regulations provide economic incentives to conserve they could be expected to be effective. The effect of simple regulation by prohibition depends upon the coercive power of the government and its bureaucracy. It may work well in some countries but not in others.

Faced with such a tantalizing theory as the "new" paradigm one immediate reaction is to test it against other relevant data using conventional econometric techniques. Unfortunately, as discussed earlier, the data needed to test these hypotheses adequately are not yet available. Even for the OECD countries, the data that would enable us to observe the level of per capita income above which there is a decline in environmental deterioration (at the peak in Figure 1) are not available.

Plausible air pollution data series for the U.S. for sulfur oxides, nitrogen oxides, and total suspended particulates dating from 1940 to 1990 (Council on Environmental Quality 1992) shown in Figures 3, 4 and 5 do appear to confirm the decline with increasing GNP per capita, population and urbanization, and also over time, after reaching a peak. This unfortunately is not good enough to demonstrate that the theory holds for countries undergoing the four processes mentioned above. However, even without the detailed time series, one can appeal to studies such as Ridker's (1972) and compare predictions with outcomes. For example, the actual 1990 emission of air pollutants in the U.S. is in each case less than one-third of Ridker's year 2000 predictions assuming no changes in environmental policy, and in each case also less than the corresponding 1970 emission levels. Similarly in terms of resource usage, despite a 25% increase in population and an almost 50% increase in per capita income since 1970, consumption of all primary minerals and energy resources are less than one-third of the forecast year 2000 amounts. The use of most minerals in 1990 was at or below their 1970 use. Total energy use had increased to only 118% of the 1970 level, compared with the 260% to 690% predicted by the year 2000. This level of misprediction over such a short time period should be viewed as cautionary advice to would-be modelers. At the very least it casts doubt on drawing too much from the elasticities computed by Ridker and reported in Tables 2 and 3. At best, it is a corroboration of the new paradigm showing decreases in resource use over time with a proportionally larger decrease in the emission of pollutants. In this interpretation, it shows both a decoupling of resource use and per capita income and also a decoupling between resource use and emissions of pollutants.

Emissions and Time
U.S. 1940–1990

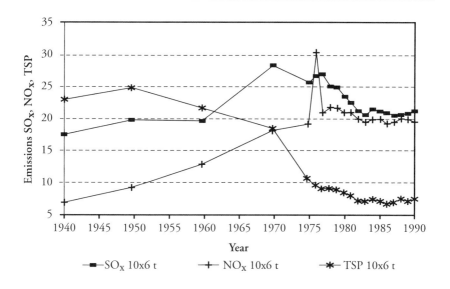

Emissions and Time
U.S. 1940–1990

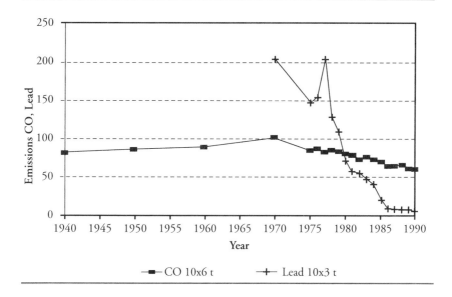

FIGURE 5

Emissions and Per Capita Income
U.S. 1940–1990

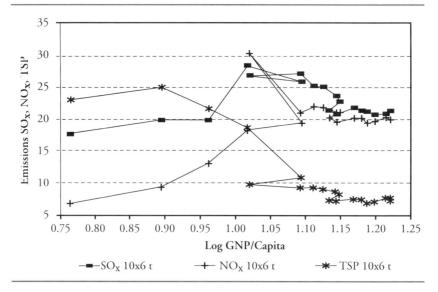

Implications of the New Paradigm

Whether to base environmental policies on the new paradigm is still in doubt. If the new paradigm were accepted, the implications would be profound. It would vindicate the current strategy of many developing countries, and of the multilateral and bilateral development agencies. According to the paradigm, the way to achieve environmental improvement is to go through a period of rapid economic growth, which even though it could be environmentally destructive, would lead both to a backlash by the citizenry against the environmental insults, and to the accumulation of economic assets which would enable the environmental insults to be cleaned-up. With higher levels of disposable income, preferences change to more leisure time activities which typically require a clean ambient environment. Citizens can afford the time and money to organize to pressure government to bring about environmental improvements which are now much more valued. One way to characterize this is:

> Left to themselves, the forces of economic development and citizen's participation in democratic nation-states will eventually lead to demands for improvement of the human environment and ultimately the ecosystem. (Rogers 1992)

CO2 Emissions Per Capita
1950–1989 (Boden, et al, 1991)

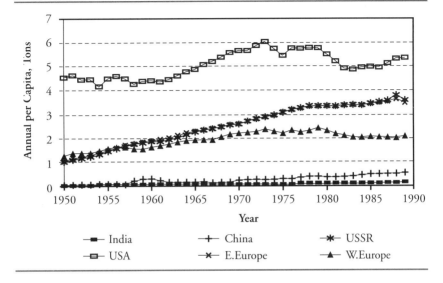

The word "democratic" is key in this characterization. The experiences of the former Soviet Union and the Eastern Bloc countries underline the importance of allowing for citizen participation (via outrage) to fuel the movement for environmental repair.

Ecologists may be uneasy with this approach which they would characterize as "overshoot and correction." They may ask "What if in the process of overshoot we cause irreparable damage?" The answer to this is best seen in the "environmental assaults of the third kind"—the global issues of carbon dioxide build-up, global warming, and loss of ozone. In the resolution of the first two cases the paradigm appears to work well. As resource constraints become more obvious, the efficiency of combustion is improved, processes are changed, and the economies essentially decouple themselves from carbon dioxide production. This has happened without need for citizens' outrage. Figure 6 shows estimates of the total emissions of carbon dioxide for India, China, the former USSR, the U.S., Eastern Europe and Western Europe from 1950 to 1989. The figure suggests that Western Europe has already passed through its peak, that the U.S. passed through a peak in the late 1970s and bounced up again in the late 1980s, and that the remaining countries had not yet reached their peaks. Questions remain as to whether production will continue to decline in Western Europe and the U.S. or stabilize at current

FIGURE 7

Total CO2 Emissions
1950–1989 (Boden, et al, 1991)

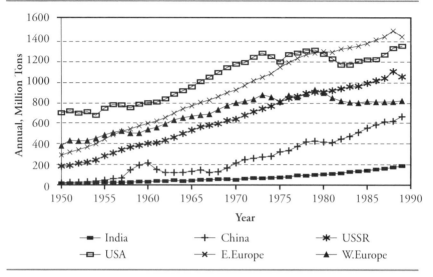

levels. Figure 7, which shows the same data, may hold the answers. All the industrialized countries appear to show a peaking of per capita emissions, earlier rather than later for the western countries. Also, given the very high per capita emissions in the U.S., it appears that large reductions to bring it down to the level of Western Europe are possible if the current efforts to increase fuel taxes are successful.

Even in the case of ozone layer depletion, the awareness of the problem (identified by the developed countries) has led to rapid acceptance of measures to reduce the offending chemicals. Some would argue that in this case we came too close to irreparable damage. But, in fact, the new paradigm also seemed to work quite well in this situation.

The paradigm may not work in cases such as that of ecosystem sustainability. It is unlikely that anyone will seriously attempt to restore the millions of hectares of wetlands that have been filled and drained around the globe. Therefore we will not be able to sustain the aquatic wildfowl populations at their historic levels. The future also looks bleak for the Royal Bengal Tiger! Indeed, a doubling of the global population will put major strains upon all forms of wild terrestrial species. Short of saving the gene pool in some sort of zoo system, it is unlikely that many species will survive in the wild—unless the human species moves down the food chain and returns to largely vegetarian diets and to the sun for its energy requirements.

Unfortunately, the paradigm puts the responsibility increasingly upon the poor countries. They are the ones to the left of the peak in Figure 1, and they will ultimately determine the outcome. For example, as Figures 6 and 7 demonstrate, the huge populations of China and India make large and increasing contributions to the global level of emissions despite low per capita emissions. Some projections identify them as the leading producers of CO_2 by the early part of next century. If global warming turns out to be as serious a problem as some commentators suggest, then China and India will be in the thick of attempts to reduce emissions at a time when their economies are not well-prepared to face the costs. Rich countries could share the costs and international efforts along these lines have already begun (see Jhirad and Tavoulareas, 1993).

Finally, if the paradigm is correct, is there a particular level of GNP per capita at which we can expect the environment to begin to improve? This is at about $3000 per capita in Figure 2, where the environmental deterioration is reversed and starts to improve. If it were approximately that amount, then we could expect to see a turn around on environmental improvement coming in about 20 years in Asia and Latin America, and in 40 to 50 years in Africa. These may seem to be long time periods, but in terms of ecological processes they are relatively short. The future does not look as bleak as many predictions suggest. If population growth is slowed, then the income levels at which environmental protection becomes more important will be reached sooner. Hence, this complacent prediction does not justify any slackening in the efforts to limit population and to speed development. What it does imply is that economic development can make an important contribution to improving environmental quality. Shafik and Bandyopadhyay (1992) stated it slightly differently:

> The evidence suggests that it is possible to "grow out of" some environmental problems. But there is nothing automatic about this—policies and investments must be made to reduce degradation. The econometric results presented here indicate that most countries do choose to adopt those policy changes and to make those investments, reflecting their assessments of the evolution of the benefits and costs to environmental policy (Shafik and Bandyopadhyay 1992, 23).

REFERENCES

Asian Development Bank. *Key Dedicators of Developing Asian and Pacific Countries.* Manila: Economic and Development Resource Center, Asian Development Bank, July, 1991.

_____. *Economic Policies for Suitable Development.* Manila: Asian Development Bank, 1990.

_____. *Asian Development Outlook 1991.* Manila, 1991.

Australian Delegation to the World Population Conference. *World Population Conference 1974, Report of the Australian Delegation.* Australia: Australian Delegation to the World Population Conference, 1974.

Bernstam, M. S. "Wealth of Nations and the Environment." In *Resources, Environment, and Population.* Supplement to *Population and Development Review*, Vol. 16:333–373, edited by M. S. Bernstam and K. Davis. 1990.

Christmas, J., and C. de Rooy. "The Decade and Beyond." *Water International* 16 (1991):127–34.

Council on Environmental Quality. *Environmental Quality: 22nd Annual Report.* Washington, D.C., March, 1992.

Department of Technical Co-operation for Development. *Report of the International Conference on Population 1984* (Doc#: E/CONF76/19). New York: United Nations, 1984 (ILS HB 849 I536 1984).

ESCAP. *State of the Environment in Asia and the Pacific 1990.* Bangkok: United Nations, Economic and Social Commission for Asia and the Pacific, 1992.

Gardner, R. N. *The United Nations and the Population Problem: The Proceedings of an International Conference*, May 1973. In preparation for the 1974 World Population Conference. New York: The Institute on Man and Science, 1973.

Isard, W., et al. *Ecologic-economic Analysis for Regional Development: Some Initial Explorations with Particular Reference to Recreational Resource Use and Environmental Planning.* New York: Free Press, 1972.

Jhirad, D. J., and E. S. Tavoulareas. "International Promotion and Financing of Thermally Efficient and Environmentally Acceptable Clean Coal Technologies." Paper presented at the Second World Coal Institute Conference, London, England, March 24–26, 1993.

Jorgensen, D. W., and P. J. Wilcoxen. *Reducing U.S. Carbon Dioxide Emissions: The Cost of Different Goals.* Discussion Paper Series No 1575. Harvard University: Harvard Institute of Economic Research, October, 1991.

Keyfitz, N. "Population and Development Within the Ecosphere: One View of the Literature." *Population Index* 57, no.1 (Spring 1991):5–22.

Leontief, W. "Environmental Repercussions and Economic Structure: An Input-Output Approach." *Review of Economics and Statistics*, Vol LII, no. 3, August, 1970.

Munasinghe, M. *Water Supply Policies and Issues in Developing Countries, Natural Resources Forum.* Illinois: Butterworth & Co (Publishers) Ltd., February, 1990.

Organization for Cooperation and Economic Development, 1991. *The State of the Environment, Annual Report.* Paris.

Population Reference Bureau, Inc. *1991 World Population Data Sheet.* Population Reference Bureau, Inc., 175 Connecticut Avenue NW, Suite 520, Washington, DC 20009, 1991.

Repetto, R., et al. *Wasting Assets: Natural Resources in the National Accounts.* Washington, D.C.: World Resources Institute, 1989.

Revelle, R. *Population and Environment.* Research Paper No. 4. Harvard University: Center for Population Studies, May, 1974a.

Revelle, R. "Will the Earth's Land and Water Resources Be Sufficient for Future Populations?" In *The Population Debate: Dimensions and Perspectives*, vol. 2. Papers of the World Population Conference, Bucharest, 1974b, ST/SEA/SERA/57. New York: United Nations, Department of Economic and Social Affairs, 1975.

Ridker, R., ed. *Population, Resources, and the Environment.* The Commission on Population Growth and the American Future, US Government Printing Office, 1972.

Rogers, P. 1992. Personal comments given at the Roger Revelle Memorial Symposium on Population and the Environment, Cambridge.

Ruttan, V. *Sustainable Growth in Agricultural Production: Poetry, Policy, and Science*, Staff Paper, 91–47. University of Minnesota: Department of Agricultural and Applied Economics, 1991.

Shafik, N., and S. Bandyopadhyay. *Economic Growth and Environmental Quality: Time Series and Cross-Country Evidence.* Background paper for the World Development Report 1992, Policy Research Working Papers, WPS 904. The World Bank, Washington, DC, June, 1992.

Slade, M. E. "Natural Resources, Population Growth, and Economic Well-Being." In *Population Growth and Economic Development: Issues and Evidence*, Ch. 9, D. Gale Johnson and R. D. Lee., eds. Wisconsin: University of Wisconsin Press, 1987.

United Nations. *World Population Trends and Policies: 1981 Monitoring Report Volume I: Population Trends* (Doc#: ST/ESA/SERA/79). New York: United Nations, 1982.

United Nations. *World Population Trends and Policies: 1981 Monitoring Report Volume II: Population Policies* (Doc#: ST/ESA/SERA/79/Add1). New York: United Nations, 1982.

UNDP. *Human Development Report 1992.* Oxford University Press, 1992.

UNEP. *United Nations Environment Programme: Environmental Data Report.* Blackwell, 1987.

UNEP. *United Nations Environment Programme: Environmental Data Report.* 2nd edition, Blackwell, 1989.

U.S. Environmental Protection Agency. 1991. "National Air Pollutant Estimates, 1940–1989," *Report EPA-450/4-91-004.* March, Research Triangle Park, NC.

Water International. New Delhi Statement: Some for All Rather than More for Some. *Water International* 16 (1991): 115–20.

World Commission on Environment and Development. *Our Common Future.* Oxford University Press, 1987.

World Bank. *World Development Report, 1992: Development and Environment.* Oxford University Press, 1992.

World Resources Institute. *World Resources 1986.* Basic Books, 1986.

World Resources Institute. *World Resources 1987.* Basic Books, 1987.

World Resources Institute. *World Resources 1988–89.* Basic Books, 1988.

World Resources Institute. *World Resources 1990–91.* Oxford University Press, 1990.

World Resources Institute. *World Resources 1992–93.* Oxford University Press, 1992.

NOTES

1. This paper has benefited from the helpful comments of Professor Robert Dorfman, Stephen Marglin, Christian Dufournaud, Rodney Tyers, Rosemarie Rogers and Mr. Nagaraja Rao Harshadeep.

III

POPULATION SCIENCES

6

THE "REVELLE REPORT" ON RAPID POPULATION GROWTH: A 1971 DIAGNOSIS AND PRESCRIPTION REVISITED

By Paul Demeny

The sharp acceleration of global population growth following World War II was largely unanticipated by demographers and other social scientists. By the late 1950s, however, fairly general awareness of this development had emerged and generated intense analytic work and policy discussions. This led to the launching of programs with international assistance aimed at lowering the rate of population growth in high fertility countries. Private organizations pioneered in this endeavor, but by the mid-1960s the official foreign aid organizations of several Western governments were authorized to extend assistance for population programs in developing countries and they soon became by far the most important players in the field. The year 1965 marked the entry into this field by the U.S. government, the largest and most active international donor.

Concern with the consequences of rapid population growth had an obvious focus: high fertility. Of the two other direct determinants of country-level growth—international migration and mortality—the former was of minuscule importance in most high fertility countries. As for mortality, declining death rates *were* the prime cause of accelerating population growth. But the mortality decline that had already taken place was, of course, uniformly welcomed, and further enhancing gains in survival was a natural objective not only of individuals but also of public policy.

The choice of the best means for pursuing the goal of lowered fertility was, arguably, less obvious. Prior to World War II, many countries experienced a decline in fertility, sometimes at a remarkably rapid pace, without deliberate inducement by governments. In those countries, to the extent that governments took any interest in aggregate fertility, their intervention was pronatalist. Thus, among those concerned with high fertility, few considered the Western record of fertility decline as offering relevant lessons for shaping antinatalist policy. Besides, the suddenness of the demographic growth surge after World War II seemed to call for rapid engagement in resolute and effective action. Modern means of birth control and new contraceptive devices coming online held out the promise of a technological solution. They also offered a tangible core around which programs could be organized, and provided a compelling rationale for the need for international assistance. The technology in question, after all, *was* both sophisticated and Western; virtually unknown and unavailable in the developing world. Although relatively low-cost, it was not easily affordable to governments with meager resources. A workable mode of delivering it to potential users was uncharted territory, but was obviously akin to the operation of the successful public health programs in which international assistance has played, and was then still playing, an important and highly effective role. Finally, a series of attitude surveys undertaken in high fertility countries suggested a strong latent demand for birth control. Internationally assisted family planning programs—programs offering the modern methods for birth control at nominal or no cost to those who want them—thus became the centerpiece and, apart from activities ancillary to them, were in most cases the sole instrument of efforts aimed at controlling high fertility.

As demonstrated by their success in convincing initially reluctant donor and recipient governments to take up a new endeavor that was demanding in terms of financial and administrative resources, advocates of this novel policy approach had good arguments. Still, at the time of the birth of large-scale family planning programs, the intellectual underpinnings of the policy were less than satisfactory. A report of the U.S. National Academy of Sciences (NAS) prepared in 1963, just two years before the launching of the US Agency for International Development's (USAID) population activities, illustrates this clearly. Addressing the "problem of uncontrolled population growth," the report called it "one of the most critical issues of our time," but in its recommendations it focused on the need for better understanding and more research and training. It also recommended the establishment of another NAS committee to deal with the issue (National Academy of Sciences 1963).

The Academy's exhortation for more research and analysis, reinforced by similar sentiments among other influential observers, elicited a salutary response. Within a few years following that call, primarily through support from private foundations and endowments, a number of research institutions and programs were set up at prestigious American universities. (For an account, see Caldwell and Caldwell 1985.) Inter alia, the Harvard Center for Population Studies was born, and placed under the leadership of the eminent scientist Roger Revelle. In 1967, a meeting of the directors of seven U.S. centers for population research concluded that "not enough is known about the economic and social effects of rapid population growth" (National Academy of Sciences 1971a,v). This was a notable claim, as the presumed deleterious effects of such growth on development constituted the main rationale on which the by then active U.S. foreign assistance program in population was said to be grounded. "Research," the directors complained, "on the social, economic, political, and educational consequences of high and sometimes rising birth rates, falling mortality, differing age patterns in changing societies, and what a policymaker can do about these phenomena, has lagged sorely behind the research on demographic and contraceptive aspects of the problems" (ibid.).

Action to remedy this situation soon followed under Revelle's leadership. In the same year, 1967, he took the question to the then Administrator of USAID, William S. Gaud, who in turn asked the National Academy of Sciences to study the questions on which knowledge was found to be deficient. Dr. Frederick Seitz, president of the Academy, appointed a 12-member special study committee, chaired by Roger Revelle, to examine the issues involved, commission contributions from individual scholars and submit a report to the Academy and to USAID.

The results, prepared mostly in the summers of the following two years, appeared in a 700-page volume four years later under the title *Rapid Population Growth: Consequences and Implications* (National Academy of Sciences 1971a). The work bears the marks of Revelle's intellectual leadership and the effectiveness of his kind but firm pressures on invited authors to deliver. His demanding guidance, as I fondly remember, was most effective in the give and take of the long, periodic working sessions framed by the bucolic surroundings of Woods Hole.

Rapid Population Growth is composed of two parts. The second part, that provides its bulk, consists of 17 individual research papers focused on various aspects of the topic. The first 100 pages are taken up by the report of the Committee (the twelve members of which included none of the invited contributors except T. W. Schultz) for which Revelle, as chair, was primarily responsible. This section, which was also published as a separate book in a paperback edition (National Academy of Sciences 1971b) and was promptly dubbed and has ever since been commonly referred to as the "Revelle Report," bears the title, "Summary and Recommendations." But the label "Summary," if taken as reference only to the rest of the volume, is inaccurate. That opening section stands as a closely argued and coherent precis of the then-state-of-the-field and as a comprehensive articulation of the policy conclusions derived from it by Revelle and his fellow committee members. It is a summation of the insights of the dominant strand of American thinking during the first two postwar decades about international population issues. Offering both diagnosis of the problem of rapid population growth and a set of prescriptions for solving or mitigating that problem, it is undoubtedly Roger Revelle's most influential contribution to the field of population studies and population policy.

In what follows, with the benefit of hindsight, I will discuss some salient aspects of the diagnosis and the prescriptions the Revelle Report offered. To attempt a summary of this already complex 100-page summary in a few pages would be, of course, futile, as well as unnecessary. Even after more than two decades, this document should be familiar to everyone involved in its subject, and it should be obligatory reading for anyone newly entering the field of concern it addresses. I will conclude with some observations that perusal of the report elicits today.

Diagnosis

The Revelle Report was based on what in retrospect can be characterized as a fairly accurate assessment of the global demographic situation. Its estimate of the world population in 1970 was 3.5 billion, some 200 million below what later became the official UN estimate for that year. It is worth noting this fact, since a correct numerical grasp on global population size was then a very recent achievement. Not surprisingly, for most observers it shed a startling, and even frightening, light on what came to be popularly referred to as the "population explosion." To be sure, demographers were long aware that by historical standards global population growth was already rapid before mid-century. The average annual growth rate in the nineteenth century was slightly above one-half of one percent: it brought the world's population from an estimated 950 million in 1800 to somewhat above 1.6 billion in 1900. Two world wars notwithstanding, the average annual growth rate in the first half of the twentieth century was still higher: only slightly below nine-tenths of one percent. But, from that already rapid tempo of increase, the jump to a 2 percent rate of annual growth caught all experts unprepared.

Thus, in 1945, Frank Notestein, then the most prominent figure among American demographers, foresaw a year 2000 population of some 3 billion. In retrospect, we know that that figure was surpassed as early as 1960 (Notestein 1945). In 1949, assessing the global food situation in the remainder of the century, M. K. Bennett (director of Stanford's Food Research Institute), still took that long-range forecast as valid (Bennett 1949). In 1954, at the Rome World Population Conference, the first major international scientific meeting in demography in the post-World War II period, E. F. Schumacher (later the author of *Small is Beautiful*) was aware that the global demographic picture had significantly changed. Assessing the adequacy of future energy supplies, he took 3 billion as the appropriate estimate not for the year 2000 but for 1980 (Schumacher 1955). This was a projection that ventured only a quarter century forward, yet it still turned out to be in error by an amount roughly equivalent to the total world population at the beginning of the twentieth century. Some 11 years later, in 1965, at the next World Population Conference, held in Belgrade, the prominent Soviet demographer, A. Y. Boyarsky, was more cautious. He offered his year 2000 global population forecast defined by a range: somewhere between 4.2 billion and 5.0 billion (Boyarsky 1967). But this generous spread failed to guard against error. The actual size of the world's population surpassed the lower-end figure by 1977; the upper-end estimate was surpassed just 10 years later.

Such failures of global foresight (of which, looking back to the 1950s and 1960s, many more could be cited) reflect the accumulation of errors in estimating country population trends. At least until around 1960, demographers' assessments of population growth prospects displayed a consistently conservative bias, particularly with respect to developing countries. For example, in his encyclopedic and authoritative study of the Indian subcontinent, Kingsley Davis estimated that the ultimate population of the area, to be reached sometime in the twenty-first century, may reach 700 million (Davis 1951). In fact, the population of the Indian subcontinent—post-partition India, Pakistan, and Bangladesh—today exceeds that perceived upper limit by some 450 million and the World Bank's current projections anticipate eventual stabilization at a level exceeding 2.5 billion.

In similar fashion, around 1950, demographers professed to perceive a huge potential for rapid population growth in China. But the quantitative estimate of that potential reflected a deep pessimism concerning the severity of the environmental constraints thought to limit China's population growth. The various estimates of demographic growth projected to be experienced in that country between 1950 and 1980 ranged from 65 million to 140 million. The actual increase during that period, despite a devastating famine that in the late 1950s resulted in about 30 million excess deaths and a birth deficit of similar magnitude, was about 440 million.

In contrast to the examples of misjudgment just cited, assessment of the global demographic situation in the Revelle Commission's work, as I noted above, no longer exhibited a failure to appreciate the historically unprecedented scale of the ongoing demographic expansion. In a number of respects, in fact, its bias was in the opposite direction. Although the report duly noted then already observed rapid declines in seven "poor countries" between 1960 and 1967, and appreciable declines in seven other developing countries, it strongly suggested that rates of population growth in the developing areas would continue to increase. Accordingly it offered a year 2000 global population forecast of 7.5 billion as plausible and 7 billion as "the best estimate now possible." These estimates are not as egregiously erroneous as the examples cited above, but the direction of the bias is significant. In retrospect, we know that the rate of both global population growth and the growth of the developing world as a whole reached their all-time high about the time when the report was composed, and a slow but unmistakable downward trend of these rates had begun by the time the report was published.

The report missed signaling even the likelihood of this coming historic turnaround. In a recent survey of fertility trends that highlights declines in the total fertility rate since the mid-1960s, Ronald Freedman and Ann Blanc say that "very few observers were predicting such declines in 1965" (1992, 44). The dictum would seem to fit the Revelle Report's stance, but not a number of contemporary forecasts. For example, the influential volume on Third World population and development trends prepared in the mid-1950s at INED, the French demographic research institute, very much anticipated coming generalized fertility declines (Balandier 1956). Even more apposite is a comparison with the authoritative UN population projections of the time. For example, the UN's most commonly cited "medium" projection of the year 2000 global population was 6.1 billion in the 1963 assessment of the world demographic situation and 6.3 billion according to the 1973 assessment. (The corresponding current UN estimate, assessed in 1992, projects a 2000 population of 6.2 billion [United Nations 1992].) Both of these early UN forecasts reflect anticipation of the onset and/or steady continuation of fertility declines affecting much of the population of the developing world during the remainder of the century.

The Revelle Committee's skepticism on this score reflects an assumption that underlies much of the report's analysis: the assumption that patterns of demographic behavior observed in Western populations, including during their premodern stage of development, have little relevance to the understanding of demographic processes in the contemporary developing world. But such putative Western exceptionalism has been clearly contradicted by Japanese demographic history and, by the mid-1960s, also by the recorded experience of those fourteen non-Western countries referred to above. Remarkably, around the time the NAS study was in progress, Revelle was also immersed in a project pulling together studies on demographic history. In assessing these studies, which he edited, Revelle commented that "less developed countries are much more like Japan and the countries of Europe two hundred years ago than those countries today." He further noted the "marked differences in levels of marital fertility and, hence, presumably in the degree of fertility control within marriage among the regions of Europe and at different times in the same region" (Revelle 1972, 14). The point of delving into that experience was (in the words of the British sociologist-demographer David Glass, who was Revelle's coeditor), that understanding processes of change in preindustrial historical populations can throw light upon prospects in developing societies. Disappointingly, this interest left no mark in the Revelle Committee's report, either in its diagnoses or in its prescriptions.

The absence of discussion of the prospects of a spontaneous demographic adjustment process in the report is doubly surprising in view of the by then well-recognized pattern of exceptionally rapid economic growth present in countries comprising the majority of the developing region. If markets were allowed to operate so that incentives affecting individual behavior would be set right, could not material advance along the earlier Western patterns be expected? And with even modest material advance, amplified by increasing exposure to the outside world through the new technologies of transportation and communication, would not fertility adjustment follow? Must the decline of fertility—an eventual necessity if mortality was to be kept low—trace a qualitatively different path in the Third World than the one that was followed, with differing time lags but with accelerating tempo, in the countries that achieved low levels of fertility and were now called "developed"? If birth rates dropped by 50 percent during the two decades between the two world wars in relatively backward Bulgaria, could that not also be expected to happen in late twentieth century Brazil or Burma?

The Revelle Committee did not consider such questions. What it clearly and correctly foresaw was the coming massive upsurge of population size in the developing countries during the last three decades of the twentieth century under *any* plausible scenario for fertility trends. Its main diagnostic effort was therefore focused on the consequences of rapid demographic growth for development. The assessment it offered presents a sober view. Improvements in health apart, rapid population growth is depicted as a hindrance, and often a major brake on development. The report emphasizes that "technical potentialities exist, not only to feed all human beings, but greatly to improve the quality of human diets." And it notes that "the natural resources available to present technology are sufficient to allow a vast improvement in the standard of living of all the people who will inhabit the earth 20 to 30 years from now." The adverse effect of rapid demographic growth is not that it makes development impossible but that it makes improvements in the human condition more difficult to achieve than would be the case under conditions characterized by slower growth. Guarded optimism that is warranted for the earth as a whole, of course, may be less easily argued for the less well-placed components of the global population.

The report traces the deleterious effects of rapid population growth on six specific areas. They are the economic, the social, and the political effects, and the consequences for education, health, welfare, child development and the environment. Discussion of each of these issues judiciously summarizes consensus findings and findings on which there was, at the time, near agreement, albeit with a bent on highlighting problems and passing over areas of ignorance or areas of population-insensitivity. Such areas were, arguably, numerous and significant. The task, as the Committee apparently saw it, was to state findings backed by a reasonable consensus. Future research, vigorously urged by the report, was to dispel ignorance on the various socioeconomic consequences of population growth and to keep abreast of unfolding developments.

The Revelle Report provided what was seen in the following 10 to 15 years as a solid rationale for the international population activities of the U.S. government. Secure in that knowledge, the funding agencies ignored the Committee's call for continued research. Support for follow-up studies on the consequences of population growth was scanty. Detailed analyses of particular country situations were practically nonexistent, although for the purposes of policymaking such studies could have been especially valuable. In academia, an increasingly influential revisionist current emerged, as is ably chronicled in Kelley (1988), eroding the standard economic rationale for population policies set forth by the Revelle Report. Such revisionism was epitomized by Julian Simon's 1981 book and, most notably, by a 1986 report of the National Academy of Sciences that reexamined the issues discussed by its predecessor committee fifteen years earlier (National Research Council 1986). Qualitatively, these two NAS reports do not differ greatly; both tend to see population growth as a drag on development. The quantitative significance assigned to these effects in the second report, however, is strikingly different. In the 1986 report's scale-insensitive neoclassical frame of reference, the impact of population growth on development is found to be typically modest and often negligible. As a result, by the late 1980s, official justification for the international population activities of USAID increasingly dropped reliance on the deleterious economic effects of rapid population growth, invoking instead the argument that population programs satisfy real needs of their clients. Thus, the type of programmatic action that was endorsed and articulated by the Revelle Committee's report—organized family planning programs—is being relegated to the status of one of the many competing government-run or government-supported social service programs. In such a competition, family planning programs are not especially well-placed, either as a matter of empirical fact or as a matter of intrinsic merit. This is not surprising: unlike many other services (for example vaccination programs), family planning services deliver goods that can be, if of inferior quality, home-produced. But, arguably, this shift in the supporting rationale for population programs is explained more by an inadequate intellectual grasp on the nature of the problem of rapid population growth than by intrinsic irrelevance of such growth in shaping the prospects for development. The diagnosis offered by the Revelle Committee on this score needs to be reexamined. Very likely, reexamination would reconfirm its essential soundness.

Prescription

The centerpiece of the Revelle Committee's recommendations for fertility-reducing policies was family planning programs. "It is today possible," the report stated, "for a full range of acceptable, easily used, and effective means of preventing births to be provided by governments to all persons of reproductive age, if necessary at nominal or no cost. Steps and actions can be taken to foster broad social legitimization and support of birth control, including, when circumstances permit, medically safe abortions." This was, of course, hardly a novel idea: save for the last clause it was the program urged by private organizations since at least the early 1950s, and espoused by the

U.S. government, and later by multilateral agencies, notably the United Nations Fund for Population Activities, as an object of international assistance for developing countries in the 1960s. The report provided additional support for what was received official policy.

The report stated firmly that acceptance of the services offered by family planning programs must be voluntary, but noted that freedom to make reproductive decisions must be tempered by concern for the rights and interests of others. "One of the most difficult of population questions," said the report, "relates to designing and justifying governmental policies and procedures to accomplish this end." The report went on briefly to outline a number of suggestions on this score—an agenda for sending the right signals to individuals about social expectations on matters of reproduction.

But the recommendations of the report concerning this second, demand-oriented leg of fertility policy, i.e., recommendations designed to affect social attitudes and provide economic incentives that would generate desires for lower fertility on the micro-level, lacked conviction and clear articulation. Most certainly, they lacked programmatic appeal for those in charge of population policy. In retrospect we know that these recommendations produced little or no tangible results. The reasons for this are not difficult to discern. The business of development in the postwar world came to be conceived as a planned, government-directed enterprise, in which specialized departments were in charge of assigned program areas. Programs that could be packaged around a distinct and tangible technology had a strong advantage in claiming attention and support. Success could be further assured by the creation of organizational structures and cadres needed to carry out a program, as these became reliable sources of advocacy and political clout, sustaining the programs of which they were in charge.

The involvement of foreign assistance for such programs worked to reinforce the tendency for central direction and sectoral packaging of activities—a tendency that also matched the natural inclination of aid recipient governments. Disbursing foreign aid required local officials who had administrative responsibility for programs and could be held accountable. Family planning activities easily fit into a sectoral division of labor, natural to planners; the task of program administrators, once given their organizational slot and assignment, was to get their fair share of the action in competing with other claims for limited resources. Proliferation of an increasing variety of sectoral endeavors, each with its own well-articulated agenda aimed at satisfying some "unmet" but worthy need, was a tendency built into this model of development policy. So was the over-commitment of resources, leading to underfunded programs that chronically promised more than they could deliver. More often than not, family planning programs did not fare well in the game.

The encouragement given by aid donors, de facto although not always in words, to this developmental model is a peculiar feature of postwar international relations. Western success in rising from backwardness to relative affluence (and, pari passu, from high to low fertility) was not achieved along such a path; indeed, donors had no home-based experience whatsoever with many of the programs they sought to foster and partly finance in countries with governments possessing very modest domestic resources, including organizational and administrative capacity. They did have experience with successful recipes for development, experience also highly relevant with respect to influencing demographic behavior, but these were in the domain of constitutional design: a social technology not considered fit for export through foreign assistance.

The Revelle Report did not delve into the difficult issues adumbrated in the preceding paragraphs. It is doubtful that its recommendations aimed at fostering micro-level demand for lower fertility were offered with much hope for effective follow-up. A committee addressing a problem area that, to begin with, was designated as a specialized domain of a subdepartment within a U.S. government agency, was ill-placed to take issue with the accepted general strategy of international development activities.

As it turned out, the report's main population program recommendation—support for family planning services—certified its value by survival. It had both success in attracting resources and disappointment in being chronically underfunded in relation to its perceived potential. As to policies that would have increased demand for the modern contraceptive services offered by family planning programs or would have prompted individuals to control their fertility by means of traditional methods that proved effective in the demographic transition of the West, they sometimes received lip-service, often not even that. But the rapid economic and social changes that most developing countries experienced in the post–World War II decades did produce shifts in demographically-relevant incentive structures that helped to nudge fertility lower even when such shifts were not, as they could have been, part of the design of social policy. Schools may be a free government service, centrally financed, but keeping children in them can become costly. Health services can be declared a basic human right to which everyone is entitled free of charge, but keeping children healthy may call for increasing expenditure of parental resources.

How to apportion the credit for the declining birth rates (that now have been experienced by the majority of the developing world's population) between these two forces—supply of the means of birth control, and shifts toward lower desired fertility unwittingly engineered by economic and social change—remains an unresolved and controversial question.

Since family planning programs, in various shapes and sizes, are now ubiquitous, often any decline tends to be routinely credited, in toto, to the impact of the preexisting family planning program. Such an interpretation often is a reflection of simple analytic innocence. Since the dominant, although by no means only, proximate determinant of fertility change is increased use of contraception, and since (when programs are active) the devices used are supplied, not surprisingly, by the programs that provide them free of charge, the cause-effect relation seems straightforward enough. But some specialists, pointing to the many examples of fertility declines in the past that occurred (sometimes with great speed, in the absence of any government program and, to boot, prior to the advent of modern contraceptive technology) discount program effects drastically. The most influential expert discussions of the issue tend to avoid such extremes of interpretation and claim, if less than convincingly, a middle ground (see, for example, Phillips and Ross 1992). Whatever the actual effect is, it is clear that a large, and arguably the dominant chunk of the action belongs to the factor called, opaquely, "development," but meaning shifts in microlevel incentives that generate demand for lower fertility.

One would think that this state of affairs would be reflected in the balance of the attention that is devoted to these two pillars of fertility policy. As is well known to any observer of the population policy scene, this is not the case. Preoccupation with family planning programs absorbs the lion's share not only of the budgetary and physical resources allotted to the "population sector"—that is fair enough, given the difference between the technologies of the policy approaches that are called for—but it also claims much of the analytic and research capacity at the disposal of decisionmakers in this area of public concern.

Rereading the Revelle Report is a reminder that this is a wholly anomalous situation. Another reminder is reading current pronouncements of prominent public figures about the subject of international population policy. Even when coming from widely different points of the ideological spectrum, the lopsided emphasis on the supply side is apparent (see, for example, Nixon 1992, and Gore 1992). Implicitly they all endorse what may be called, unkindly, the stupid peasant theory of demographic behavior: people have as many children as they do, not because they want them but because they don't know how to prevent births.

The recently issued joint statement (the first of its kind) of the Royal Society of London and the U.S. National Academy of Sciences provides an even clearer illustration of the imbalance. The statement, entitled *Population Growth, Resource Consumption, and a Sustainable World* (Royal Society 1992), takes what may be fairly labeled an alarmist view of population growth. (It would seem that the authors of the statement are unaware of the U.S. Academy's 1986 report on the subject that set forth quite a different position.) But the conception reflected in the statement of what may contribute to a solution of that alarming problem is remarkably narrow. "Why is population growth rapid?", the statement asks. Answer: because of "large amounts of unwanted childbearing." And what can scientific research do to mitigate the problem? One single answer is given, as far as population is concerned: "development of new generations of safe, easy to use, and effective contraceptive agents and devices." This diagnosis and this prescription reflect a profound misreading of the central lessons of the development experience of the now industrialized countries; of the factors that made the West rich and that generated its fertility transition. They do a disservice to Third World economic, social and demographic development. The geopolitical changes that transformed the international scene during the last three years call into question some of the basic assumptions underlying the received policy wisdom on development. In a world contemplating another doubling of its population during the next century, a new Revellian undertaking to review and reappraise international population policy is urgently needed.

References

Balandier, G., ed. *Le 'tiers monde': sous-développement et développement.* Paris: Presses Universitaires de France, 1956.

Bennett, M. K. "Population and food supply: The current scare." *Scientific Monthly* 63, no.1, 17–26 (1949).

Boyarsky, A. Y. "A contribution to the problem of the world population in the year 2000." United Nations, *World Population Conference, 1965,* vol. 2, 5–12. New York: United Nations, 1967.

Caldwell, J. and P. Caldwell. *Limiting Population Growth and the Ford Foundation Contribution.* London and Dover, NH: Frances Pinter, 1985.

Davis, K. *The Population of India and Pakistan.* Princeton: Princeton University Press, 1951.

Freedman, R., and A. K. Blanc. "Fertility transition: an update." *International Family Planning Perspectives* 18, no.2, 44–50, 72. (1992).

Gore, A. *Earth in the Balance: Ecology and the Human Spirit.* New York: Houghton Mifflin Company, 1992.

Kelley, A. C. "Economic consequences of population change in the third world." *Journal of Economic Literature* 26, no.4, 1685–1728 (1988).

National Academy of Sciences. *The Growth of World Population.* Washington, DC: National Academy of Sciences, National Research Council, 1963.

———. Rapid Population Growth: Consequences and Policy Implications. Baltimore: Johns Hopkins Press for the National Academy of Sciences, 1971a [xxii+691p].

———. Rapid Population Growth: Consequences and Policy Implications: Summary and Recommendations. Baltimore: Johns Hopkins Press for the National Academy of Sciences, 1971b [xi+105p].

National Research Council. *Population Growth and Economic Development: Policy Questions.* Washington, D.C.: National Academy Press, 1986.

Nixon, R. *Seize the Moment.* New York: Simon and Schuster, 1992.

Notestein, F.W. "Population—The long view." In *Food for the World*, T.W. Schultz, ed. Chicago: University of Chicago Press, 36–57, 1945.

Phillips, J. F. and J. A. Ross, eds. *Family Planning Programmes and Fertility.* Oxford: Oxford University Press, 1992.

Revelle, R. "Introduction." In *Population and Social Change* edited by D. V. Glass and R. R. Revelle. London: Edward Arnold, 13–21, 1972.

Royal Society [of London and the U.S. National Academy of Sciences]. *Population Growth, Resource Consumption, and a Sustainable World.* (Joint statement by the two Institutions), 1992.

Schumacher, E. F. "Population in relation to the development of energy from coal." United Nations, *Proceedings of the World Population Conference 1954*, vol. 5, 149–164. New York: United Nations, 1955.

Simon, J. L. *The Ultimate Resource.* Princeton: Princeton University Press, 1981.

United Nations. *World Population Prospects: The 1992 Revision.* 1992 (Mimeographed).

7

MORTALITY AND THE FATE OF COMMUNIST STATES

By Nicholas Eberstadt

Introduction

Roger Revelle touched the lives of a great number of people over his long and distinguished career. This inspiring teacher also changed the lives of quite a few students. I am one of them.

As fate would have it, my very first class as a newcomer to Harvard College in Fall 1973 was Natural Sciences 118, Roger Revelle's panoramic introduction to the global dynamics of population, natural resources and the environment. It was a *tour de force*. I was hooked.

Roger Revelle's own work only tangentially touched upon the problems of the Soviet Bloc. The sorts of research and policy dialogue that he nurtured so famously in so many countries were simply not well-suited for Communist environments. But I would like to think that if he had lived another 20 years, Roger Revelle would have trained his intellect on the problems of the former Warsaw Pact region in the same way that he focused on the South Asian subcontinent in the 1960s, 1970s and 1980s. And I would like to think that he would have wanted his students to raise some of the questions that are discussed in the following essay.

The crisis and collapse of Communist rule in Eastern Europe and the Soviet Union between 1989 and 1991 was, arguably, one of the defining moments of modern history. It has also proved to be a moment of truth for the large and established community of Western specialists—within government, the academy and private institutes—who have made it their purpose to study the Soviet bloc countries. For with a sudden flash, the weaknesses and shortcomings of a generation of extensive studies were exposed and glaringly illuminated.

One may begin with what did *not* happen. Students of the Warsaw Pact region provided virtually no forewarning of the convulsions that were coming to the states they studied. Quite the contrary: the demise of Warsaw Pact Communism seemed to take the field as a whole by almost complete surprise. Though a sizable literature on the Warsaw Pact states was published in the generation before the "revolutions of 1989," one would be hard pressed to locate more than a handful of items that would seem to anticipate the fact (much less the timing) of this dramatic political departure from the Eurasian stage.[1]

Prediction of political outcomes, it must be said, remains rather more of an art than a science. Premonitions about impending political changes may speak more to intuition than to positive analysis; it may be unreasonable to fault a specific discipline for failing to announce an advent almost no one was expecting. More troubling, from a methodological standpoint, than the unpredicted collapse of Soviet bloc Communism itself has been the light this collapse has newly thrown on Western studies of the quantitative performance of those economies.

Of course, the closed nature of those societies and the peculiar nature of their economic organization, posed obvious and formidable obstacles to any effort at measuring true patterns of production, consumption and growth under central planning. But it is equally true that Western researchers devoted formidable resources to developing just such measurements. In fact, by such criteria as expense, duration or technical manpower absorbed, the U.S. intelligence community's postwar quest to describe and model the Soviet economy was in all likelihood the largest social science research project ever undertaken.

With the crumbling of the Berlin Wall and the "end of the Cold War," it now seems virtually incontestable that this corpus of quantitative studies, for all their apparent rigor, was for the most part off the mark. As late as 1987, for example, the U.S. Central Intelligence Agency (CIA 1987) was estimating per capita GNP to be somewhat higher in East Germany than in West Germany. As every German taxpayer is today all too aware, the actual level of per capita output in what have become the "New Federal States" was in reality vastly lower than in the rest of Germany on the eve of that nation's reunification.[2] This signal misreading of Communist economic performance was by no means an isolated incident. Comparable overestimates of economic performance by the CIA may be adduced for every other country in the Soviet bloc. And the CIA was hardly alone in overestimating productivity, living standards and economic growth within the Soviet bloc during the "Cold War." In a recent defense of his analysts, the Bush Administration's Director of Central Intelligence pointedly (and correctly) observed that many independent scholars and specialists had taken the CIA's old estimates to task for purportedly underestimating the economic performance of Warsaw Pact countries (Gates 1992).

My purpose in reciting these particulars is not to criticize the work of those in the West who strove so diligently to understand the workings of a system, and a way of life, with which they were fundamentally unfamiliar. It is rather to suggest, with all the benefits of hindsight, that our understanding of the Soviet bloc might have been significantly enhanced if students of those countries had paid a little more attention to their demographic trends. More specifically, I wish to suggest that valuable and telling insights into the social, economic, and even political circumstances of these countries could have been gleaned from analysis of their death rates: their schedules of mortality, and their reported patterns of cause of death.

Such an application of demographic technique, one hopes, would have been gratifying to Roger Revelle. It responds, after all, to an invitation he extended a quarter of a century ago. In offering the results of that now-classic project on "Historical Population Studies" to the reader in 1968, Revelle also framed an agenda for future research. "The new science of historical demography," he observed, "has devoted almost all of its efforts to the determinants of population change, and very few to an examination of its consequences." Among the specifics that "we do not understand," in his enumeration, were "the consequences of greater longevity on allocation of resources and the distribution of political power" (Revelle 1968).

What we have witnessed in the Soviet bloc since 1968, however, is an unprecedented demographic phenomenon that could scarcely be imagined at the time: long-term stagnation—even decline—in life expectancy among a group of industrialized societies that were not at war. This was a major and highly visible event, rife with consequence and implication. Some of those consequences and implications will be explored in this paper.

The paper proceeds through four sections. The first reviews the anomalous history of mortality trends in Eastern Europe and the USSR between the end of the Second World War and the "end of the Cold War." The second draws inferences about economic performance in those countries from their mortality trends. The third examines some characteristic differences in mortality trends between those areas in which Communist rule has recently collapsed and those in which it continues, and speculates about the significance of the distinction. The final section discusses the significance of current mortality trends for post-Communist societies, especially as they pertain to the prospective transition to a stable economic and political order.

I

For the Soviet Union and the Eastern European countries over which it gained mastery during and immediately after World War II, the "Cold War" began with an explosion of health progress. Improvements in mortality, of course, had been underway in these areas before the advent of Marxist-Leninist rule, and had proceeded even during the crisis years for interwar capitalism: between 1929/32 and 1937, for example, Czechoslovakia's life tables recorded an increase in expectation of life at birth of about three years for males and three-and-a-half for females (United Nations 1968). Nevertheless, the pace of mortality decline under Red Army Socialism was noteworthy. Even after recovery to prewar levels had been attained, the tempo continued to be brisk. It was not idle conceit to assert during those years, as the Soviet bloc governments did, that they were outperforming their capitalist rivals in the field of public health.

A few summary figures will lay out the contrast. Between the early 1950s and the early 1960s, according to recent United Nations Population Division estimates, infant mortality dropped by nearly half in Warsaw Pact Europe, and by over half in the USSR (United Nations 1991). This was a much more rapid pace of progress than in Western Europe as a whole, or even in Western Europe's Mediterranean regions (which were arguably more comparable to the Balkans or Ukraine than Switzerland or Sweden would have been). Over that same decade, life expectancy at birth in Warsaw Pact Europe is estimated to have risen by over five years for males, and by nearly six-and-a-half for females; in the USSR the corresponding increments are estimated at five-and-a-half and four-and-a-half years, respectively. Improvements in "imperialist countries" were not nearly so dramatic. In West Germany, for example, the overall increase in life expectancy at birth between 1950/55 and 1960/65 was about two-and-a-half years; in the United States, it was barely one year. Among the OECD countries, Japan could claim to match the average of Warsaw Pact Europe's increments, but her performance was exceptional for a grouping in which she was clearly an atypical member (ibid.).

By the mid-1960s, mortality levels in the Warsaw Pact region were very nearly as low as those of the advanced capitalist countries. By the UN's estimate, less than a year separated overall levels of life expectancy at birth in the U.S. and the USSR; the differential between OECD Europe and Eastern Europe was apparently only very slightly greater. If trends were extrapolated only a few years into the future, the Soviet bloc could be seen catching up with, and then surpassing, advanced Western countries by this important and politically significant measure.

That general crossover point, of course, was never reached. Instead, the Warsaw Pact countries collectively, and most unexpectedly, entered a new era: one of stagnation and deterioration in overall health conditions. The sudden slow-down is highlighted by the official life tables of the countries in question. (Despite the problems with some of these life tables, largely attendant upon underreporting of infant deaths; their results are informative.) According to its life tables, life expectancy at birth in Poland rose by less than two years between the mid-1960s and the late 1980s—this apparently was the region's star performer. Overall increases of less than one year over this same generation were estimated by Bulgaria, Czechoslovakia, and Romania; Hungary's life expectancy was basically stationary over the period as a whole; and life expectancy at birth is officially reported to have fallen in the USSR. Worse still was the situation for men: according to these life tables, East Germany enjoyed an increase in life expectancy at birth for males of about one year over these decades, and this was the best of the group. Czechoslovakia's male life expectancy in 1988 is placed at the same level as in 1964, while Bulgaria, Hungary, Poland, Romania and the USSR all report long-term declines. The USSR also reports a slight long-term decline in female life expectancy at birth.[3] Strikingly, the slowdown, and reversal, of health progress within the Soviet bloc coincided with an acceleration in OECD countries. Between the mid-1960s and the late 1980s, according to the latest UN estimates, life expectancy for both sexes rose by an average of about four years in Western Europe, and by about five years in the United States (United Nations 1992).

How is the virtual cessation of overall health progress within the Soviet bloc to be explained? The phenomenon may be better understood by examining component parts. In most of the Warsaw Pact region, infant mortality rates apparently continued their declines through the 1960s, 1970s, and 1980s, although at a slower pace than in the early postwar period. (There are strong indications that Soviet infant mortality actually increased over some portion of the past generation; owing to the poor state of the relevant Soviet data, however, the debate about this possibility has not yet been settled.)[4]

TABLE I

Expectation of Life at Age 1 and Age 30: Warsaw Pact Region, c.1965–c.1989

Country	Life Expectation at 1 Year of Age (years)		Life Expectation at Age 30 (years)	
	Male	Female	Male	Female
Bulgaria				
1965/67	70.28	73.81	43.06	45.99
1987/89	68.42	74.64	40.87	46.53
increment	−1.86	+0.83	−2.19	+0.54
Czechoslovakia				
1964	68.44	73.96	41.15	45.84
1988	67.70	75.07	39.73	46.62
increment	−0.74	+1.11	−1.42	+0.78
East Germany				
1967–68	69.77	74.70	42.46	46.70
1987–88	69.53	75.46	41.67	47.08
increment	−0.24	+0.76	−0.79	+0.38
Hungary				
1964	69.08	73.45	41.74	45.45
1989	65.58	73.86	37.84	45.55
increment	−3.50	+0.41	−3.90	+0.10
Poland				
1965/66	68.98	74.43	41.68	46.46
1988	67.37	75.70	39.60	47.29
increment	−1.61	+1.27	−2.08	+0.83
Romania				
1963	69.36	72.96	42.66	45.65
1987/89	67.53	73.13	40.55	45.56
increment	-1.83	+0.17	−2.11	-0.09
Soviet Union				
1965/66	—68[1,2]—		— 45[1]—	
1986/87	—67.2[1,2]—		—43.5[1]—	
increment	*negative*		*at least one year*	
1958/59	65.62	73.07	39.51	46.13
1989/90	65.29	74.43	38.62	46.78
increment	−0.33	+1.36	−0.89	+0.65

Notes: [1] = life expectancy for both sexes; [2] = life expectancy at age 5

Sources: For USSR 1958/59 and 1986/87 = USSR State Statistical Committee, *Tablitsy Smertnosti i Ozhidaemoy Prodalzhitel'nosti Zhizni Naseleniya* (Moscow: Goskomstat, 1989). All others, United Nations, *Demographic Yearbook* (New York: UN Department of International Economic and Social Affairs), various issues.

Mortality data tend to be more comprehensive and reliable for those over the age of one than for infants. They are worth examining, not least for this reason. Table 1 presents official life table estimates of life expectation at age one for the Soviet bloc countries for the mid-1960s and the late 1980s. A striking pattern emerges from these data. In five of the seven countries, combined male and female life expectancy for the noninfant population registers at least a slight decline. The situation is even starker by age 30: in six of the seven countries, life expectancy for adults fell over these decades, slight increases in female life expectancy being more than offset by the fall in life expectancy for males. In less than a quarter of a century, Bulgarian and Polish life expectancy for men at age 30 dropped by more than two years; in Hungary the drop was closer to four years. The Soviet Union, for its part, never released any life tables for its population for the mid-1960s. Official tables for the late 1950s and the late 1980s, however, report male life expectancy at age one, and life expectancy at age 30 for the two sexes together, to have been lower in 1989/90 than they were over thirty years earlier!

Age-specific death rates can cast further light on the deterioration of adult health within the Soviet bloc (see Table 2). Between the mid-1960s and the late 1980s, all seven countries reported rising death rates for at least some of their adult male cohorts. Some of these increases were little short of astonishing: in Hungary, for example, death rates for men in their forties doubled between 1966 and 1989. In most of these countries, women in various adult cohorts experienced at least some slight declines in age-specific mortality, although broad increases in female mortality were evident in both Hungary and the Soviet Union. Although health trends were, by this measure, arguably unfavorable in all these countries for all adult cohorts, they were especially bad for persons in their forties and fifties.

The health situation—call it a health crisis—in the Soviet bloc countries during their last generation under Communism was without historical precedent or contemporary parallel. Mortality decline in Western countries, it is true, has been neither smooth nor uninterrupted; various countries—including the Netherlands, Sweden and the United States—have reported drops in life expectancy at birth for their male population at some juncture during the postwar period. These drops, however, have been slight and temporary, whereas the rise in death rates for broad groups within Warsaw Pact countries has been major, and sustained over the course of decades.

What accounts for these extraordinary trends? A fully satisfactory answer to this question must await further interdisciplinary study. Some preliminary indications, however, may be drawn from data on reported cause of death in these countries. These data must be used with care, for they are shaped by an unavoidable element of subjectivity under the best of conditions, and the best of conditions did not obtain in the statistical offices of Warsaw Pact states. Cause-of-death data, nevertheless, can speak directly to the proximate reasons for reduced life expectation, and may be broadly suggestive of the underlying factors driving the decline.

Changes in Age Specific Death Rates for Cohorts Aged 30–69: Warsaw Pact Region, c.1965–c.1989 (percent)

Country and Sex	Cohort Age							
Males	30/34	35/39	40/44	45/49	50/54	55/59	60/64	65/69
Bulgaria (1966–89)	+19	+32	+62	+70	+56	+47	−16	+14
Czechoslovakia (1965–89)	−5	+8	+19	+40	+33	+29	+15	+6
East Germany (1965–88)	−5	−8	−5	+7	+3	+1	−15	−17
Hungary (1966–89)	+67	+96	+100	+131	+93	+69	+46	+25
Poland (1966–88)	+9	+17	+36	+51	+47	+38	+23	+6
Romania (1966–89)	+32	+36	+43	+61	+44	+32	+35	+15
Soviet Union (1965/66–89)	−5	0	+21	+25	+24	+25	+20	+25
Unweighted average	+17	+26	+39	+55	+43	+34	+15	+8
Females								
Bulgaria (1966–89)	−11	−15	−10	+4	−4	−4	−7	−6
Czechoslovakia (1965–89)	−13	−23	−14	−9	−10	0	−3	−9
East Germany (1965–88)	−12	−15	−14	−12	−12	−10	−16	−17
Hungary (1966–89)	+33	+26	+26	+33	+23	+22	+7	−2
Poland (1965–88)	−27	−25	−9	−9	−2	−1	−3	−14
Romania (1966–89)	−8	+13	+4	−3	0	−2	−3	−3
Soviet Union (1965/66–89)	−21	−17	−4	−3	+2	+11	+4	+19
Unweighted average	−8	−8	−3	0	−1	+2	−3	−5

Note: All changes rounded to the nearest percentage point. Percentages derived from sources.

Sources: For Soviet Union 1965/66: John Dutton, Jr., "Changes in Soviet Mortality Patterns, 1959–77." *Population and Development Review,* vol. 5, no. 2 (1979), pp. 276–277. All other data: United Nations, *Demographic Yearbook* (New York: UN Department of International Economic and Social Affairs), various issues.

The World Health Organization (WHO) has prepared age-standardized breakdowns of mortality rates by reported cause of death for all of its corresponding member states for the period extending from the early 1950s to the present. It offers breakdowns for the 1965/69–1989 period for four of the seven countries in the Warsaw Pact: Bulgaria, Czechoslovakia, Hungary and Poland. Some of the trends highlighted are intriguing. Levels of mortality attributed to accident and injury, for example, are consistently high among these states; Hungary—the country of the four with perhaps the best cause-of-death data—reports that age-standardized mortality from cirrhosis and chronic liver disease was, by 1987, higher for women than it had been for men only seventeen years earlier. But mortality ascribed to these causes does not trend upward with any consistency in these countries, and, in any case, cannot account for much of the overall increment in age-standardized death rates. It is, instead, deaths attributed to cardiovascular disease (CVD) that appear to have shaped these countries' age-standardized mortality trends. In all four countries, deaths attributed to CVD accounted for over half of all age-standardized mortality by the late 1980s. Moreover, all four countries are reported to have suffered huge rises in CVD mortality levels between the late 1960s and the late 1980s. Whereas mortality ascribed to CVD has declined substantially in recent decades throughout the Western industrialized world, Bulgaria reported a 52 percent increase in age-standardized CVD mortality for males between 1965/69 and 1989. Over the same period, CVD mortality seems to have risen dramatically for East bloc women as well: by over 16 percent in Bulgaria and 20 percent in Poland, to select two of the more arresting examples. In proximate terms, the explosive rise in CVD-attributed mortality seems to account fully for the rise in age-standardized mortality in these Soviet bloc countries.[5] Diverging levels of CVD mortality, moreover, seem to account for most of the divergence in overall age-standardized mortality rates over the past generation between Western and Eastern European countries.

The underlying factors contributing to this rise, of course, are more difficult to identify than the proximate ones. Deaths from cardiovascular disease are commonly associated with a variety of specific behavioral or lifestyle characteristics, including heavy smoking, heavy drinking, poor diet, lack of exercise and psychological stress or emotional strain.[6] There is ample evidence to suggest that Warsaw Pact populations may have been increasingly exposed to such "risk factors" as the era of "detente" progressed.

But such behavioral indications, in a sense, beg the question, for they skirt an obvious etiological issue: not only is the phenomenon of secular increases in mortality in industrialized societies at peace a new one, the patterns accounting for this rise are also unprecedented. Higher levels of general mortality are typically associated with higher levels of death from infectious and parasitic disease, whereas the Warsaw Pact group's path back to higher mortality was paved by increases in deaths attributed to chronic, noncommunicable causes. The rise, and its pattern, was unique to populations living under Warsaw Pact Communism, and indeed apparently common to all of them, despite their manifest differences in language, culture and levels of material attainment. By these forensics, one might surely be drawn to inquire whether the health problems evidenced by the Warsaw Pact countries were not, in some fundamental sense, systemic. Moreover, in an age when health progress is all but taken for granted, and when scientific, technical and administrative advances have made it possible to attain given levels of mortality at ever-lower income levels, an inability of a particular set of governments to prevent severe long-term declines in health conditions for broad segments of its populations is surely suggestive of a systemic crisis.

II

During the Cold War decades, Western efforts to assess the Soviet bloc economies and to measure their performance were hampered not only by secrecy and mutual mistrust, but by features characteristic of Soviet-style command planning.

For one thing, the incentive structure in the Soviet-type planning rewarded overstatement of results at all levels, including the very highest. As Jan Winiecki (1986) once aptly observed, the "law of equal cheating" does not obtain in such a milieu. Even figures for physical output were routinely padded and exaggerated, albeit by varying margins across countries, industries and time.

Yet even if perfectly accurate time series data had been available for all items of physical output, Western analysts would still have faced a second problem: valuing the goods and services produced in a way that would make them comparable with output from a market-oriented economy. Ingenious attempts to translate a price structure set by the state into one reflecting scarcity costs were devised: most importantly, the "adjusted factor cost" method pioneered by Abram Bergson and his colleagues in the 1950s.[7] But the problem of finding a common valuation process for systems which allocated resources by such fundamentally different principles could neither be finessed, nor ultimately solved.

For all these difficulties, economic estimates for the Warsaw Pact group were produced, and internally consistent time series were developed to trace their performance. The most authoritative of these time series, published by the CIA, did indicate a fairly steady slowdown in the tempo of economic growth for the region as a whole between the mid-1960s and the late 1980s. At the same time, it suggested quite considerable economic progress. For the 24-year span 1966–1989, for example, the CIA estimated per capita output in Warsaw Pact Europe to have risen by over half; per capita growth for the period as a whole was said to average 1.8 percent a year. By this reading, the region's growth rate would have been lower than the European Community's (2.4 percent for those same years), but the ostensible gap was not dramatic. Specific comparisons, moreover, painted a more favorable picture of the Warsaw Pact's performance against its rivals. The CIA's estimates of per capita growth for the U.S. and the USSR for these years, for example, were virtually identical (1.9 percent per annum), and East Germany's per capita growth rate was placed slightly ahead of West Germany's (2.7 vs. 2.6 percent). On the eve of the "revolutions of 1989," furthermore, CIA estimates indicated that the Warsaw Pact economies had attained fairly high levels of productivity. For 1988, per capita output for the USSR was placed at over three-fifths of the West German level. Per capita output in Czechoslovakia was placed at 78 percent of the level for the Netherlands, and per capita output in East Germany was estimated to be virtually the same as for the European Community as a whole (CIA 1988).

Plausible as such numbers may have seemed when juxtaposed solely against one another, they would seem suspicious—indeed anomalous—if held next to mortality data. Mortality statistics, for example, would immediately seem to call into question the proposition that per capita output in Eastern Germany had reached the level of Western Europe's by the late 1980s. After all, in 1989 WHO's age-standardized death rate for males was 26 percent higher for East Germany than for the countries of Western Europe. For females, East Germany's age-standardized death rate was fully 32 percent higher than Western Europe's. Indeed, by this measure of mortality, death rates in Eastern Germany were actually higher than in such places as Argentina, Chile, Uruguay or Venezuela (WHO 1991)!

Such discrepancies in mortality levels are pertinent to economic performance in a number of respects. General levels of mortality bear more than a passing relation to labor productivity, which in turn establishes constraints on a population's level of per capita output. Mortality levels, moreover, are directly related to a population's living standards, which are in turn related to its level of per capita consumption.[8]

The relationship between mortality and economic performance, of course, is neither tight nor entirely mechanical. A country's level of per capita output is determined by more than just its supply of "human capital." Human capital, for its part, is a complex fabric of many strands, of which health is but one, and for which mortality rates may not always provide a satisfactory proxy.[9] Finally, the very fact that fairly low levels of mortality can today be purchased in some low-income countries (such as Sri Lanka or China) should qualify generalizations about the relationship between overall levels of per capita consumption and overall levels of mortality. With such caveats in mind, we may nonetheless be able to read an economic significance from the exceptional mortality trends of the Warsaw Pact region.

Consider once more the comparison of *fin de regime* East Germany with the countries of the European Community on the one hand, and with selected Latin American countries on the other. Is it conceivable that a country with general levels of mortality so much higher than the EC's average could manage to attain the EC's level of per capita output? In theory, yes, but only under three specific, and highly restrictive, conditions.[10] Equivalent levels of output could coincide with such different mortality levels if: 1) resource allocation were markedly more efficient in the high mortality society; 2) the high mortality society enjoyed a markedly superior endowment of such factors of production as capital or technology; and/or 3) the high mortality society mobilized its labor force in a way that allowed it to elicit vastly greater hours of work from its typical resident.[11]

Were these conditions satisfied by East Germany in the period leading up to unification? Soviet-type economies may be good at various tasks, but allocative efficiency was never one of these.[12] Thus, Condition 1 does not obtain. The same may be said for Condition 2: even during the "Cold War," it was no secret that East bloc industry lagged considerably behind the EC in most fields of production with respect to deployed technology and capital stocks.[13] What about Condition 3? Soviet-style systems do seem to be effective in achieving high rates of labor force participation: census data, for example, indicated that over 54 percent of East Germany's population was economically active in the 1980s, as against West Germany's 48 percent.[14] But East Germany's extensive employment strategy entailed the induction of more marginal laborers into the workforce; consequently, the average number of hours worked per week was reportedly lower than in West Germany (under 36 vs. over 40 in 1988 in the nonagricultural sectors) (International Labor Office 1991). Total hours worked per year, on a population-equivalent basis, appears to have been only slightly higher in East Germany than in West Germany (or by extension in other Western European countries). Insofar as none of the conditions adduced appears actually to have obtained, one would conclude from these mortality differentials, in the absence of other evidence to the contrary, that per capita output in East Germany was actually substantially lower than within the EC countries in the years immediately before reunification.

What of the comparison between East Germany and, say, Argentina? Is it plausible that levels of per capita consumption would be higher for the society with the higher general mortality level? Once again, the answer is: in theory, yes, but only under specific conditions. Such a paradox might be explained by peculiarities of the income distributions of the countries in question; by differences in the reach and scope of healthcare policies; or by differences in the availability and incidence of other "public consumption" goods and services, such as rationed staples, medical care or education. Yet by any of these criteria one would expect mortality levels to be lower in an Eastern European socialist economy than in a Latin American economy characterized by equivalent levels of per capita consumption. In the absence of other evidence, therefore, mortality levels would appear to indicate that East Germany's level of per capita consumption was rather more like a Latin American country's than that of a Western European country—and that it may actually have been lower than per capita consumption levels in some parts of Latin America. By way of perspective, one may note that the World Bank's "purchasing power parity" adjustments give Argentina a level of per capita output roughly one-third that of West Germany for 1985 (World Bank 1992).

This approach to mortality analysis can be extended to render a more general impression of the performance of Soviet-type economies in the generation before their demise. One instructive comparison comes from matching estimates of change in per capita output with changes in age-standardized mortality (WHO "European Model") for males over roughly the same period. That particular measure of mortality would seem appropriate as an alternative aperture on economic progress for two reasons. First, though it is a summary measure of mortality for all age groups, the model age-structure upon which it rests is heavily weighted toward persons of working ages, and is therefore sensitive to their mortality trends. Second, despite doctrinally stipulated equality of the sexes, labor force participation rates within the Soviet bloc were always higher for men than for women; moreover, Soviet bloc men tended to be disproportionately represented in higher-pay sectors. For these reasons, Soviet bloc output might be expected to be affected more by changes in male mortality than changes in female mortality.

Table 3 presents a match-up of CIA and WHO estimates for Soviet bloc countries and selected states from Western Europe. In all of the Soviet bloc countries, rising male mortality levels coincide with what are calculated to be substantial gains in per capita output. The situation in these four countries is contrasted with four of Western Europe's "slow growers": the Netherlands, Sweden, Switzerland, and the United Kingdom. Over the past generation, faster-growing Western European economies have also been characterized by a somewhat faster pace of change in age-standardized mortality; mortality change in the slow growers is less than the Western European average. Even so, these four countries exhibit an entirely different pattern of mortality change from the four Warsaw Pact countries, despite purportedly similar magnitudes of per capita growth. Where the Warsaw Pact group's mortality rates all rise, Western Europe's "slow growers" all register declines. In view of this radical difference, is it really possible that the Eastern and Western

TABLE 3

CIA Estimates of Changes in Per Capita GNP vs. WHO Estimates of Changes in Age-Standardized Male Mortality: Selected Warsaw Pact And Western European Countries c.1965–c.1989

Country	Estimated Changes in Capita GNP 1966–89 (percent)	Estimated Changes in Age-Per Standardized Male Mortality, 1965/69–89 (percent)
Bulgaria	+61.2	+13.4
Czechoslovakia	+62.7	+1.8
Hungary	+57.3	+12.5
Poland	+62.3	+7.9
unweighted average	+60.9	+8.9
Netherlands	+63.4	−11.5
Sweden[1]	+61.1	−13.3
Switzerland	+51.6	−27.1
United Kingdom	+63.0	−25.1
unweighted average	+59.8	−19.4

Notes: WHO age standardization is for its "European Model" population. [1] = 1988

Sources: Derived from U.S. Central Intelligence Agency, Handbook of Economic Statistics: 1980 edition, p. 29; 1990 edition, p. 44; World Health Organization, World Health Statistics Annual: 1988 edition, Table 12; 1990 edition, Table 10; 1991 edition, Table 11.

European countries in Table 3 would actually have experienced similar per capita growth rates over the generation in question? Under certain conditions, possibly so—if, for example, the Soviet bloc had enjoyed a clear and overriding advantage with regard to technological innovation. But as has been noted, no such advantage was in evidence then, or can be seen in retrospect.

As a final comparison, one may match CIA (per GNP capita) and WHO (age-standardized mortality) estimates for the Soviet bloc and for Latin America for the year 1989 (see Table 4). According to the CIA's assessment, none of these Soviet bloc countries had a level of per capita output nearly so low as the most affluent of these Latin American nations; as a group, their level of per capita GNP was said to be over three times as high as those Latin American countries listed. Those same Latin American societies, however, reported substantially lower levels of age-standardized mortality. Of the entire Warsaw Pact grouping in 1989, in fact, only East Germany could have passed for an advanced Latin American society on the basis of its age-standardized mortality figures. These mortality data do not offer a precisely

TABLE 4

CIA Estimates of Per Capita GNP vs. WHO Estimates of Total Age-Standardized Mortality: Warsaw Pact Countries and Selected Latin American Countries c.1989

Country	GNP ($1989)	Age-Standardized mortality, 1989 (deaths per 100,000)
Bulgaria	5690	1141.0
Czechoslovakia	7900	1158.0
East Germany	9670	1014.7
Hungary	6090	1229.6
Poland	4560	1118.7
Romania	3440	1240.5[1]
Soviet Union	9230	1159.9[1]
unweighted average	6654	1151.8
Argentina	2250	1043.7[2]
Chile	1880	969.0[2]
Mexico	2340	1026.3[3]
Venezuela	2100	1003.8
unweighted average	2143	1010.7
Ratio of unweighted averages (Latin America = 100)	310	114

Notes: WHO age-standardization is for "European Model" population. [1] = 1988; [2] = 1987; [3] = 1986

Sources: Derived from U.S. Central Intelligence Agency, Handbook of Economic Statistics 1990 edition, pp. 30–34; World Health Statistics Annual, 1990 edition (Table 10), 1991 edition (Table 11).

calibrated adjustment of official Western estimates of Soviet bloc economic performance, but they appear to provide a strong implicit challenge to such figures, and thereby may be seen as serving something of a corrective function.

Even among market-oriented societies, international economic comparisons remain a complex and exacting business, for reasons both practical and theoretical. It should be no surprise that opportunities for mismeasurement were greater still when surveying the economies of the Soviet bloc. It may have been too much to hope for a single, unambiguous statistical account of the performance of economies so very different from our own.

Even so, the simple device of inspecting mortality rates might have indicated a great deal about economic performance in these countries. In the absence of countervailing evidence, they would have suggested that long-term per capita growth was negligible, if indeed positive, between the mid-

1960s and the late 1980s; that per capita output was closer to the Latin American than the Western European level at the time of Soviet Communism's collapse; and that levels of per capita consumption in the Soviet bloc might approximate those of Latin America as well. With the benefit of hindsight, and a largely unforeseen revolution, it may now be said that such a reading would not look too far off the mark.

III

Though Communist rule has collapsed in the Warsaw Pact region, it continues in other lands: China, Cuba, North Korea and Vietnam among them. This partial collapse of a once-global political and economic system poses an obvious question: why did some Marxist-Leninist regimes shake and fall in 1989–91, whereas others managed to weather the storm? The question may seem most appropriate for the historian or the student of international affairs, but it can also be framed in demographic terms. For one may wonder: is it entirely a coincidence that the governments which vanished during this crisis of international Communism had all witnessed long-term health reversals among broad segments of the populations under their control, while all the governments that endured had supervised populations characterized by general and continuing mortality improvements?

The contrast in mortality trends of now-defunct and still surviving Communist governments could hardly be more vivid. Comparison of these trends, however, is not a straightforward proposition. Vital registration data, for the most part, are rather less comprehensive in these surviving states than they were within the Warsaw Pact. (Cuba is the exception to this generalization, although even there questions remain as to the quality of its infant mortality data (Hill 1983). For China, North Korea and Vietnam, demographic trends had to be reconstructed on the basis of census returns and/or incomplete registration system data; such reconstructions do not permit more exacting or specific analyses of mortality conditions (e.g. year-to-year changes in age-specific death rates). They do, however, provide reasonably reliable estimates of long-term changes in general mortality levels. In particular, it is possible to estimate changes in life expectancy at birth over the past generation for these three countries (see Table 5).

In all four countries, gains in life expectancy are quite substantial. Whereas life expectancy for males stagnated, or declined, in the Warsaw Pact countries between the 1960s and the 1980s, it is reported to have increased in Cuba by over eight years, and is estimated to have risen by nearly 14 years in North Korea. China and Vietnam appear to have enjoyed gains in life expectancy at birth of nearly a decade for both males and females over the

TABLE 5

Officially Claimed or Independently Reconstructed Changes in Life Expectancy at Birth: Surviving Communist Regimes, c.1965–c.1989

Country and Year	Life Expectancy at Birth for Males (year)	Life Expectancy at Birth for Females (year)
Cuba		
1965	65.4	67.2
1983/84	72.7	76.1
increment	+7.3	+8.9
China		
1965	56.6	57.9
1989	66.2	67.0
increment	+9.6	+9.1
North Korea		
1965	51.0	57.1
1987	64.7	71.0
increment	+13.7	+13.9
Vietnam		
1965	49.7	57.0
1989	62.0	65.9
increment	+12.3	+8.9

Notes: Cuban data taken from official life tables. Estimates for China, North Korea and Vietnam are reconstructions based on census and/or registration data.

Sources:
Cuba: United Nations, *Levels and Trends in Mortality Since 1950* (New York: UN Department of International Economic Aid and Social Affairs, 1982), p. 174; Demographic Yearbook 1990. (New York: UN Department of International Economic and Social Affairs, 1992), p. 490.

China and Vietnam: unpublished estimates, U.S. Bureau of the Census, Center for International Research.

North Korea: Nicholas Eberstadt and Judith Banister, *The Population of North Korea* (Berkeley, CA: University of California, Institute of East Asian Studies, 1992), pp. 108–9.

course of that same generation. By any historical measure, progress in reducing mortality in all these countries over the past decades would arguably qualify as rapid.

This is not to say that health progress in the surviving Communist states has been steady and consistent. China's "Great Leap Forward" (1957–58) brought on a demographic catastrophe, in which tens of millions perished and life expectancy plummeted. More recently, Beijing's antinatal population policies have been associated with infanticide, and rising mortality, for baby girls; by some estimates these increases were for a while sufficiently consequential to reduce life expectancy at birth for females (Banister 1987). Such reversals, however, were tied to specific political campaigns. With the relaxation or reversal of the afflicting policy, mortality reductions in both cases resumed their downward trend under the same standing government.

Is there a political significance to this broad distinction in mortality trends within the former Communist world? I believe there may well be, and that the distinction may be useful to understanding the very different fates of these two sets of regimes.

Let me be clear. I do not propose to replace the Marxist notion of historical materialism with determinism of a demographic variety, or to deny the significance of individual actions and discrete decisions within the great play of history. The collapse of Soviet bloc Communism can be traced through a progression of specific events that was in no sense "historically inevitable": the accession of Mikhail Gorbachev; Moscow's decision to abide by the results of Poland's 1989 elections; Budapest's announcement later that year that it would permit East German "vacationers" to use Hungary as a transit stop on the way to West Germany; the failed coup in Moscow in August 1991. I wish instead to underscore the fact that, in our era at least, long-term rises in mortality are fraught with an unavoidable political significance. Whatever else they may portend, secular increases in mortality may today be read as an indicator of fragility for a regime, or an entire system.

The systemic political significance of the Warsaw Pact region's long-term health trends may be better appreciated by comparing their circumstances with the experiences of the Latin American and Caribbean region during the 1980s. For that area, after all, the 1980s were a decade of economic crisis and social reversals; of debt default and "structural adjustment." Whatever the inexactitudes of the calculations, the World Bank now estimates per capita output in the region as a whole to have been lower in real terms in 1990 than it had been in 1980; in countries such as Argentina, Panama, Peru and Venezuela, corresponding estimates suggest that per capita GNP was over 15 percent lower at the end of the decade than at its start (World Bank 1992). Local austerity measures typically targeted public consumption—the "social safety net,"—and most countries in the region could point to cutbacks in public expenditures on income support, education and health. (Even before these cutbacks, Latin America's "social safety nets" were not considered famously sturdy.)

Despite these various shocks and setbacks, there is to date no evidence of pervasive increases in mortality in the Latin American and Caribbean region. To be sure, the vital registration systems in many of these countries—including some of the ones apparently hardest hit—were poorly developed, and would not have been capable of providing an immediate representation of mortality reversals. Nevertheless, demographers who have examined mortality trends in the region since the advent of "structural adjustment" have concluded—and sometimes to their own admitted surprise—that there is as yet no reason to believe that the direction or even the tempo of mortality change has been affected by this great depression.[15]

To judge by their mortality trends, the population of the Warsaw Pact region, in the generation leading up to the fall of their Communist regimes, was suffering through a much more drastic crisis than the one that jolted Latin America in the 1980s. We may lack a compass adequate to the task of charting economic change in these Communist societies; even so, we know that governments in other parts of the world and in other times have been able to forestall mortality reversal during severe economic downturns and dislocations. Why were these Warsaw Pact regimes not up to the challenge?

The very fact of secular mortality increase is evidence of a serious failure in health policy. But it is suggestive of much more. Mortality conditions are affected by a constellation of social, economic and environmental factors. Education, housing conditions and environmental quality are but a few of the areas bearing upon health in which virtually all modern governments routinely intervene. In centrally planned economies, where government arrogates a more far-reaching authority over the social and economic rhythms of life, the correspondence between mortality trends and government performance is presumably all the more comprehensive and direct. For the Warsaw Pact governments, the secular rises in mortality in the populations under their supervision may have been suggestive of an inability to cope—of uncorrectable policy, administrative incapacity, or even of the erosion of the governing power of the state.

In the event, those Marxist-Leninist regimes that outlasted Warsaw Pact Communism may (individually or collectively) be overturned in the near future, or instead may prove able to hold onto power for many more years. If they fall, moreover, their demise (individual or collective) may be preceded by secular mortality decline. After all, most coups and revolutions in our century have been preceded (and foliowed) by periods of general improvement in local public health conditions. Broad rises in mortality over long periods of time are neither a necessary nor sufficient condition for the collapse of a country's government, or its political system. Such rises, however, do signify the existence of extraordinary social, economic and even political stresses. As such, they serve as markers denoting risk for the regime in question. Thus, while it was in no sense preordained that the Warsaw Pact states would be the first of the Communist governments to collapse, neither should it be surprising that things happened to turn out this way.

IV

As post-Communist societies in Europe and the former Soviet Union contemplate the course before them, they know they will be traveling in unexplored terrain. What Janos Kornai (1990) has called "the road to economic freedom" is in fact a desired destination, not a route map. Open societies, inviolable civil liberties, established and liberal legal systems and functioning market economies may be widely desired in the post-Communist region, but in general they remain distant objectives. Traversing the no-man's land between Leninist order and a secure market order confronts would-be reformers with a monumental task. Addressing the issue of privatization, Jan Winiecki (1991) has written that the challenge is to "find a way. . . that is both economically efficient and politically acceptable." The same may be said for the entire process of transition.

Students of politics and economics have been kept busy by the almost daily changes in these areas since the end of Communism. Political and economic plans or recommendations have proliferated; political and economic analyses of local and regional prospects are now something of a cottage industry. But in looking toward this uncertain future, the student of demography may have something to contribute, too. Once again, much of this contribution can draw from the examination of mortality trends.

In transitional economies, mortality trends can be seen as imposing a variety of constraints on the realm of the possible. Mortality levels and trends, for example, will have a direct bearing on the potential productivity of labor, thus on potential economic efficiency and growth. Mortality trends also bear directly upon a household's well-being, albeit in a way that does not always show up in conventional income accounts. Dan Usher (1977) and Sherwin Rosen (1988), among others, have made the argument that consumers and individuals are likely to place considerable economic value on the improvement in their own life expectancy (and similar arguments could be extended to cover other aspects of their health status). Conversely, deterioration of health status or life expectancy represents a self-evident reduction in well-being and living standards. While mortality decline may not be invested with political significance in the modern era, continued mortality rises, as we have seen, may betoken regime fragility. In transitional situations, one would interpret secular increases in mortality to presage reduced economic potential and administrative incapacity; one might further expect the phenomenon to generate populist pressures that could translate into the political realm.

With these considerations in mind, let us examine the initial indications about mortality trends in post-Communist territories.

Eastern Germany

In a way, Eastern Germany may constitute the best of all possible post-Communist worlds. Before its demise, the GDR was widely viewed as the most efficient and productive of the Communist economies. By virtue of their incorporation into the existing Federal Republic of Germany, more-over, the territories of Eastern Germany almost instantly secured established frameworks for civil and commercial law, and were subsumed into a stable and successful political economy. Eastern Germans, furthermore, have had their transition pains eased by subsidies of a magnitude unimaginable for any other post-Communist population: due to the favorable terms upon which monetary union was concluded, and the guarantees now available through the Federal Republic's welfare state, Eastern Germans have been receiving noninvestment transfers amounting (even after PPP adjustments) to about $4,000 per person per year since the end of 1989.[16] Finally, Eastern Germany enjoyed the best health conditions of any country in the Warsaw Pact; its infant mortality rate, in fact, compares quite favorably with that of the white American population.

Auspicious as such soundings may be, the contrast between health conditions in Eastern and Western Germany on the eve of unification was stark (see Table 6). By the measure of age-standardized mortality, Eastern Germany's male death rate was 22 percent higher than Western Germany's in 1989; for females the gap was 31 percent. By this measure, Eastern Germany joined Western Europe with the distinction of being its least healthy member; age-standardized mortality rates in Ireland, heretofore the highest in the EC or EFTA, were 8 percent lower for males and 6 percent lower for females than were the GDR's in 1989 (WHO 1990). The mortality differences between East and West Germany in 1989, one may note, were greatest for the cohorts of working age—not exactly an ideal circumstance for equalizing productivity across this expanse.

Nor do preliminary figures for 1990 provide occasion for satisfaction (see Table 7). By these data, age-specific mortality rates would appear to have risen for most cohorts in East Germany between 1989 and 1990. These apparent rises were recorded despite the magnitude, and broad incidence, of hard currency transfers from the Federal Republic to Eastern Germany

Mortality Differentials in Germany, 1989: Eastern German Death Rates as a Ratio of Western German Death Rates (Western Germany = 100)

Age Group	Males	Females
0	105	97
1–4	98	112
5–14	131	103
15–24	115	122
25–34	126	126
35–44	135	109
45–54	130	123
55–64	123	132
65–74	122	140
75 +	120	126
Age standardized	122	131

Note: Age standardization refers to WHO "European Model" population.

Source: Derived from World Health Organization, *World Health Statistics Annual 1990* (Geneva: WHO, 1991), 226, 228, 380.

After "Die Wende:" Reported Changes in Age-Specific Death Rates Eastern Germany, 1989–1990 (percent)

	0	1/4	5/9	10/14	15/19	20/24	25/29	30/34	35/39	40/44
Male	−2	+48	+11	+68	+44	+37	+26	+24	+24	+4
Female	+6	+23	+63	+118	+18	+12	+23	−3	+17	−2

	45/49	50/54	55/59	60/64	65/69	70/74	75/79	80/84	85/89	90+
Male	+36	+8	+3	0	+4	−15	+16	+6	+2	+7
Female	+18	0	+1	+5	+2	−18	+10	−2	−2	+2

Note: Changes rounded to the nearest percentage point.

Sources: Statistches Bundesamt, *Statistiches Jahrbach 1991 Fuer Das Vereinte Deutschland* (Wiesbaden: Metzler Poeschel Verlag, 1991), p. 87; unpublished data, Statistiches Bundesamt.

households, and despite the opportunity to avail of Western German health care. It is possible that some portion of this rise is a statistical artifact, attributable to improved coverage under the Federal Statistical Office, but the GDR's mortality data (if not its cause of death data) had been reasonably good. The apparent broad rise in mortality in Eastern Germany, one must note, would actually mark a reversal of a trend for the area. Between 1985 and 1989, age-standardized male mortality had declined by about 6 percent for males and about 8 percent for females.

Eastern Europe

1990 was the first year of the political and economic transition from Communism in Bulgaria, Czechoslovakia, Hungary and Poland. The details of their reform programs differed, as do the estimates of their respective performance. It is generally agreed that the economies of all these countries contracted dramatically during their first year of post-Communist rule, although the significance of their output declines is an issue of debate among observers.

Less open to debate are their recorded mortality trends (see Table 8). According to these figures, age-standardized mortality for males rose in all four of these countries, and overall age-standardized mortality rose in three of them. Age-standardized mortality for females rose in one of them as well: Hungary. However one interprets these numbers, it cannot be reassuring that "adjustment" and "restructuring," in contrast with the Latin American experience, seem here to be coinciding with at least an initial increase in death rates for broad population groups. One may further ponder on the coincidence that Hungary, the country of the four with the worst recorded deterioration in mortality conditions between 1989 and 1990, is also widely viewed by international investors as the country in Eastern Europe most receptive to foreign capital, and moving fastest in the transition to a market economy.

These initial upticks, of course, may in the event be followed by sustained and rapid mortality declines. At the very least, however, it is apparent that the "environment" of policy and social or economic conditions that would bring mortality rates down is not yet in place; liberation from Communist rule in and of itself, evidently, is not a sufficient condition. Moreover, even if a rapid and sustained mortality decline could immediately be arranged for these post-Communist regions, they would not reach the levels characteristic of Western Europe today for many years. If one posited a rate of decline of 2 percent a year (an exceptionally rapid rate, equivalent to that which Chile has enjoyed over the past generation), overall age-standardized mortality rates in East Germany would not match today's Western German rates for another twelve years; Hungary would not reach today's Austrian level for twenty-three years; and Ukraine would not reach today's Swiss level for twenty-four

"After the Revolution"—Age-standardized Mortality in Eastern Europe,
1989–1990 (deaths per 100,000 population)

Country and Sex	Year 1989	1990	% change
Bulgaria			
Males	1396.6	1397.3	+ 0.1
Females	917.8	913.5	− 0.5
Czechoslovakia			
Males	1522.8	1552.1	+ 1.9
Females	888.4	874.1	− 1.6
Hungary			
Males	1624.9	1670.6	+ 2.9
Females	933.4	955.0	+ 2.3
Poland			
Males	1498.0	1670.6	+ 1.2
Females	838.5	833.0	− 0.7

Note: Age-standardization is for WHO "European Model" population.

Sources: World Health Organization, *World Health Statistics Annual* (Geneva: WHO),
1990 edition, Table 10; 1991 edition, Table 11.

years. For obvious reasons, actual convergence might be expected to take
much longer.

Reflection upon mortality conditions, in sum, emphasizes the enormity
of the challenges, and potential tribulations that lie in store for post-
Communist populations. Marxist-Leninist rule may or may not ultimately
be viewed as an enormous historical detour. Those acquainted with mortality
statistics for the post-Communist regions, however, will appreciate that
repair of the damage experienced under Communism's tenure, and attainment of Western European levels of performance, should probably be viewed
as an historical process: one that may take decades, or even generations to
complete.

REFERENCES

Amann, R., and J. Cooper, eds. *Industrial Innovation in the Soviet Union.* New Haven: Yale University Press, 1982.

Amann, R., J. Cooper, and R. W. Davies, eds. *The Technological Level of Soviet Industry.* New Haven: Yale University Press, 1982.

Anderson, B. A., and B. D. Silver. "Trends in Mortality in the Soviet Population." *Soviet Economy* 6, no. 2 (1990):191–252.

Banister, J. *China's Changing Population.* Stanford, Ca: Stanford University Press, 1987.

Bergson, A. *The Real National Income of Soviet Russia Since 1928.* Cambridge, MA: Harvard University Press, 1961.

_____ "Comparative Productivity: The USSR, Eastern Europe, and the West." *American Economic Review* 77, no. 3 (1987):342–57.

Bergson, A., and H. Heymann, Jr. *Soviet National Income and Product 1949–48.* New York: Columbia University Press, 1954.

Brada, J. C. "Allocative Efficiency and the System of Economic Management in Some Socialist Countries." *Kyklos* 27, no.2 (1974):270–85.

Collier, I. L. "The Estimation of Gross Domestic Product and its Growth Rate for the German Democratic Republic." *World Bank Staff Working Papers* no. 773 (1985).

Davis, J. S. "Standards and Content of Living." *American Economic Review* 35, no. 1 (1945):1–15.

Desai, P., and R. Martin. "Efficiency Loss from Resource Allocation in Soviet Industry." *Quarterly Journal of Economics* 98, no. 3 (1983):441–56.

Dutton, J. Jr. "Changes in Soviet Mortality Patterns, 1959–77." *Population and Development Review* 5, no. 2 (1979):267–91.

Eberstadt, N. "Health and Mortality in Eastern Europe 1965–1985." *Communist Economies* 2, no. 3 (1990):347–371.

Eberstadt, N., and J. Banister. *The Population of North Korea.* Berkeley, Ca: University of California, Institute of East Asian Studies, 1992.

Gates, R. M. "CIA and the Collapse of the Soviet Union: Hit or Miss?" Speech delivered to the Foreign Policy Association, New York, May 20, 1992.

German Federal Statistical Office (Statistiches Bundesamt). *Statistiches Jahrbuch.* Wiesbaden: Metzler Poeschel Verlag, 1991 and 1992 editions.

_____ *Zur wirtschaftlichen und sozialen Lage in der neuen Bundeslaendern.* Special edition, April 1993.

Gomolka, S. "The Incompatibility of Socialism and Rapid Innovation." *Millenium* 13, no. 1 (1984):16–26.

Hayek, F. von, ed. *Collectivist Economic Planning*. London: George Routledge and Sons, 1936.

Hill, K. "An Evaluation of Cuban Demographic Statistics, 1938–80." In *Fertility Determinants in Cuba*, P.E. Hollerbach and S. Diaz-Briquets, eds. Washington, DC: National Academy Press, 1983.

Hill, K., and A. R. Pebley. "Child Mortality in the Developing World." *Population and Development Review* 15, no. 4 (1989):657–87.

International Labor Office. *Yearbook of Labor Statistics 1989/90*. Geneva: ILO, 1991.

Kornai, J. *The Road to Economic Freedom: Shifting from a Socialist System: The Example of Hungary*. New York: W. W. Norton, 1990.

———. *The Socialist System*. Princeton: Princeton University Press, 1992.

Labedz, L. "Small Change in Big Brother." *Survey* 120 (1984): 3.

Lipset, S. M. and G. Bence. "Anticipation of the Failure of Communism." *Theory and Society* 23 (1994).

Mises, L. von. *Socialism*. London: Jonathan Cape, 1935.

Murray, C. J. L. and L. Chen. "Understanding Morbidity Change." *Population and Development Review* 18, no. 3 (1992):481–503.

Pohl, R., ed. *Handbook of the Economy of the German Democratic Republic*. Guildford, Surrey: Saxon House, 1979.

Powell, R. P. "The Soviet Capital Stock from Census to Census." *Soviet Studies* 31, no. 1 (1979):56–75.

Revelle, R. "Introduction to the Issue—Historical Population Studies." *Daedelus* 97, no. 2 (1968):352–62.

Riley, J. C. "The Risk of Being Sick: Morbidity Trends in Four Countries." *Population and Development Review* 16, no. 3 (1990):403–32.

Rosen, S. "The Value of Changes in Life Expectancy." *Journal of Risk and Uncertainty* 1, no.3 (1988):285–304.

Siebert, H. *Das Wagnis der Einheit: Eine Wirtschftspolische Therapie*. Stuttgart: Deutsche Verlags-Anstalt, 1992.

Sinn, G. and H. W. Sinn. *Kaltstart: Volkswirtschaftliche Aspekte der deutschen Vereinigung*. Tuebingen: J.C.B.Mohr, 1991.

Smil, V. "Coronary Heart Disease, Diet, and Western Mortality." *Population and Development Review* 15, no. 3 (1989): 399–424.

United Nations. *Demographic Yearbook*. New York: UN Department of International Economic and Social Affairs, (1948–1992).

———. *Levels and Trends in Mortality Since 1950*. New York: UN Department of International Economic and Social Affairs, 1982.

———. *World Population Prospects 1990*. New York: United Nations Department of International Economic and Social Affairs, 1991.

_____. *World Population Prospects: The 1992 Revisions*. New York: Department of Economic and Social Development, forthcoming.

United States Central Intelligence Agency. *Handbook of Economic Statistics*. Washington, DC: GPO, 1980, 1988, 1990 editions.

_____. *World Factbook 1987*. Washington, DC: GPO, 1987.

Usher, D. An Imputation to the Measure of Economic Growth for Changes in Life Expectancy. In M. Moss, ed., *The Measurement of Social and Economic Performance*. Cambridge, MA: National Bureau of Economic Research, 1977, 192–226.

USSR State Statistical Committee (Goskomstat). *Tablitsy Smertnosti i Ozhidaemoy Prodalzhitel'nosti Zhizni Naseleniya*. Moscow: Goskomstat, 1989.

Winiecki, J. "Are Soviet-type Economies Entering into an Era of Long-Term Decline?" *Soviet Studies*, 38(1986):3, 325–348.

_____. Transition and the Privatization Problem. *Cato Journal*, 11(1991):2, 299–309.

World Bank. *World Development Report 1992*. New York: Oxford University Press, 1992.

World Health Organization. *World Health Statistics Annual*. Geneva: WHO, 1988, 1990, 1991 editions.

ACKNOWLEDGMENT

American Enterprise Institute and the Harvard University Center for Population and Development Studies. An earlier draft of this paper was presented at the Revelle Memorial Symposium at Harvard University in October, 1992. The author would like to thank Dr. Judith Banister of the U.S. Bureau of the Census and Dr. Gunter Brückner of the German Federal Statistical Office for kindly granting permission to use unpublished data, and Mr. Jonathan Tombes of the American Enterprise Institute for his able research assistance.

NOTES

1. This was not the first time big events had gone unanticipated. Leopold Labedz, longtime editor of the former *Survey of Soviet Studies*, has written that the only Western journal to predict Khrushchev's ouster in 1964 was *Old Moore's Almanac*—an astrological guide. See his "Small Change for Big Brother." *Survey*, (1984):120, 3. For a systematic assessment of the performance of the social sciences in this field, see Seymour Martin Lipset

and Gyorgy Bence, "Anticipation of the Failure of Communism," *Theory and Society* (forthcoming).

2. The German Federal Statistical Office, for example, estimates per capita GDP in the second half of 1990 in the "new Federal States" to have been at 30.6 percent of the Western level. Statistisches Bundesamt, *Zur wirtschaftlichen und sozialen Lage in der neuen Bundeslaendern*, Special Edition, April 1993, *127.

   That precise figure, however, should not necessarily be retrofitted to the Communist period. The collapse of Communism was attended by a massive reallocation of factors of production—and also by a drop in their utilization. But inexactitudes notwithstanding, it is clear now that per capita production was far lower in East than in West Germany in the last days of *Ostpolitik*.

3. Figures are drawn from various issues of the UN *Demographic Yearbook*.

4. For one recent view of this controversy, see Barbara A. Anderson and Brian D. Silver, "Trends in Mortality in the Soviet Population," *Soviet Economy*, 6(1990):2, 191–252.

5. For a more detailed assessment, see Nicholas Eberstadt, "Health and Mortality in Eastern Europe, 1965–1985," *Communist Economies*, 2(1990):3, 347–371.

6. For one recent interpretation of the phenomenon and its correlates, see Vaclav Smil, "Coronary Heart Disease, Diet, and Western Mortality," *Population and Development Review*, 15(1989):3, 399–424.

7. See, for example, Abram Bergson and Hans Heymann, Jr., *Soviet National Income and Product 1940–1948* (New York: Columbia University Press, 1954). The fullest exposition of the approach is Bergson's classic study, *The Real National Income of Soviet Russia Since 1928* (Cambridge, MA: Harvard University Press, 1928).

8. For a classic exposition on the distinction between living standards and consumption levels, see Joseph S. Davis, "Standards and Content of Living," *American Economic Review* 35(1945):1, 1–15.

9. The distinction between health trends and mortality trends is not to be minimized. See, for example, James C. Riley, "The Risk of Being Sick: Morbidity Trends in Four Countries," *Population and Development Review*, 16(1990):3, 403–432, and Christopher J. L. Murray and Lincoln Chen, "Understanding Morbidity Change." *Population and Development Review* 18(1992):3, 481–503. For better or worse, however, mortality data are generally more available and more reliable today than are morbidity data.

10. The following discussion is premised on the assumption that the economists under consideration operate with production functions in which output is a function of capital and labour inputs—both adjusted for quality of the stocks included—and of technical and allocative efficiency, where these are exogenous parameters. Economies of scale do not figure explicitly in the model outlined above. The discussion assumes that mortality levels do not correlate negatively with other components of "human capital stock," and that there is no correlation between mortality levels and economies of scale.

11. Conventionally, of course, one measures hours of total employment against the economically active population, not the total population. The alternate denominator is used here because we are comparing output per capita for the population as a whole, not per worker.

12. The theoretical grounds for expecting socialist economies to have systemic problems with allocative efficiency were laid out in Ludwig von Mises, *Socialism* (London: Jonathan Cape, 1935), and F. A. von Hayek, ed., *Collectivist Economic Planning* (London: George Routledge and Sons, 1936). Empirical studies of the problem as it actually exists are numerous, but one might point in particular to Josef C. Brada, "Allocative Efficiency and the System of Economic Management in Some Socialist Countries," *Kyklos* 27(1974):2, 270–85; and Padma Desai and Ricardo Martin, "Efficiency Loss from Resource Misallocation in Soviet Industry," *Quarterly Journal of Economics*, 98(1983):3, 441–56. See also Abram Bergson, "Comparative Productivity: The USSR, Eastern Europe, and the West," *American Economic Review*, 77(1987):3, 342–57, where a measure of the efficiency loss attendant upon Soviet-style socialism is computed (although this loss relates to both technical *and* allocative efficiency).

13. The literature documenting technological lag in Soviet bloc industry is extensive, but noteworthy studies include Ronald Amann, Julian Cooper and R. W. Davies, eds., *The Technological Level of Soviet Industry* (New Haven: Yale University Press, 1982); Ronald Amann and Julian Cooper, eds. *Industrial Innovation in the Soviet Union* (New Haven: Yale University Press, 1982), and Stanislaw Gomolka, "The Incompatibility of Socialism and Rapid Innovation," *Millenium*, 13(1984):1, 16–26. In the words of Janos Kornai, "To sum up, low efficiency and technological backwardness and conservatism can be attributed to the combined effects of a set of system-specific factors." Janos Kornai, *The Socialist System* (Princeton: Princeton University Press, 1992), 301.

As to general issues relating to Soviet capital stock, see Raymond P. Powell, "The Soviet Capital Stock from Census to Census," *Soviet Studies* 31(1979):1, 56–75. For 1975, Bergson estimated the USSR's per capita Gross Reproducible Capital Stock (GRCS) to be 73 percent of the U.S. level; by his estimates per capita GRCS in a sample of four socialist countries averaged 71 percent of the level in a sample of seven Western mixed-economy countries, and averaged 77 percent of the Western European level in the socialist economies analyzed. See Bergson, "Comparative Productivity," 347.

In East Germany as of 1975, the estimated nominal value of the stock of capital assets was roughly 17 percent of West Germany's—with the former measured in *Marks* and the latter in *Deutschemarks.* On a per capita basis, East Germany's estimated stock of capital assets, on a nominal basis, would have been roughly 63 percent of the West German level; insofar as the price ratio with respect to investment in 1975 was believed to be about 1.16 DM/M, this would have suggested East Germany's capital stock per capita was roughly 73 percent as great as West Germany's in 1975. See Reinhard Pohl, ed., *Handbook of the Economy of the German Democratic Republic* (Guildford, Surrey: Saxon House, 1979), 31–3, and Irwin L. Collier, "The Estimation of Gross Domestic Product and its Growth Rate for the German Democratic Republic," *World Bank Staff Working Papers*, no. 773 (1985), 27.

14. Note that we are comparing two rather different systems for measuring labor force participation; while the results cannot be presumed totally comparable, they are nonetheless illustrative.

15. For one such assessment, see Kenneth Hill and Anne R. Pebley, "Child Mortality in the Developing World," *Population and Development Review*, 15(1989):4, 657–687.

16. This is not to say that German policies pertaining to the economics of unification did not leave room for improvement. A sustained examination and economic critique of those policies may be found, for example, in Gerlinde and Hans Werner Sinn, *Kaltstart: Volkswirtschaftliche Aspekte der deutschen Vereinigung* (Tuebingen: J. C. B. Mohr, 1991), and Horst Siebert, *Das Wagnis der Einheit Eine Wirtschaftspolitische Therapie* (Stuttgart: Deutsche Verlags-Anstalt, 1992).

8

BODY WEIGHT, BODY FAT AND OVULATION: RELATION TO THE NATURAL FERTILITY OF POPULATIONS

By Rose E. Frisch

You could have heard an oceanographer drop. The occasion is the American Pediatric Society meeting in 1970. I am reporting "Height and Weight at Menarche," a paper I coauthored with Roger Revelle. " What is your background?" the chairman of the session asked. (Unspoken: "What are you and your co-author doing giving a paper here?") "I am an ex-geneticist," I replied. "And who is Roger Revelle?" (Pronounced to rhyme with "jelly.") "He's an oceanographer and director of the Harvard Population Center." Incredulity all around. Actually, it was Roger and I who were incredulous that innumerable pediatricians had studied height at menarche without ever looking at weight.

That menarche paper, published in Science in 1970, was the third in a series of papers Roger and I coauthored on the adolescent growth spurt. Our research evolved from a study of the relationship between body size and caloric supplies among diverse populations (Frisch and Revelle 1969a). We followed a serendipitous finding on body weight during the adolescent growth spurt. Revelle was intrigued by the longitudinal growth data we analyzed. We found that menarche, the onset of menstruation in girls, was strongly related to body weight (Frisch and Revelle 1969c, 1971b). We were clearly on to something exciting, and Roger made time for it amid all his other projects. When I continued the research with medical collaborators, advancing from body weight to female body fat composition, Roger was a constant source of insightful comment and support during the years he was at the Center.

It had long been known that extensive weight loss in the range of 30 percent of normal weight, observed in cases of girls suffering from anorexia nervosa, resulted in loss of reproductive ability or delay of menarche. But we discovered that loss of as little as 10 percent to 15 percent of body weight would disrupt the menstrual cycle (Frisch and McArthur 1974). After we published that paper, Roger told me that women of the hill country of India "are reported to get pregnant when the harvest comes in." Thus I made the connection to the natural fertility of populations.

Although it is commonly observed that human beings, like other mammals, grow up before they reproduce, how to define the term "grown-up," and how the completion of growth synchronizes with the ability to reproduce (Kennedy and Mitra 1963) are still controversial questions. Chronological age is normally associated with sexual maturation. However, the rate of physical growth to adulthood affects the chronological age of sexual maturation. In countries with ample food supplies and modern health measures, i.e., most Western, developed countries, girls and boys attain adult size sooner, and are bigger, at every age, than children were 100 years ago. Earlier sexual maturation for both girls and boys accompanies this more rapid physical growth.

Girls in the United States now have menarche (the first menstrual cycle) at an average age of 12.6 to 12.8 years; 100 years ago menarche did not occur until girls were 15 or 16 years old (Frisch 1972; Wyshak and Frisch 1982). Today, we have found that premenarcheal girls who run 20 miles a week or who are gymnasts or dancers, also experience late menarche, on average at 15 or 16 years, or even as late as 20 to 21 years, depending on how strict their diets are, and how much physical exercise they perform (Frisch et al. 1981).

We also found that women who have already had their first menstrual cycle can disrupt ovulation and regular menstrual cycles, hence their reproductive ability, by weight loss in the range of 10 percent to 15 percent of their normal weight for height, or increased leanness. This disruption is reversible; weight gain will restore the menstrual cycle, after varying periods of time (Frisch and McArthur 1974).

We now know that when a woman's body weight is too low, or she is too lean, the part of the brain that controls reproduction (the hypothalamus) turns off the signal necessary for activation of the pituitary gland and therefore, of the ovary. A woman therefore becomes infertile (Vigersky et al. 1977). The current increase in the number of dieting women and women athletes has made this research clinically important.

These effects of undernutrition and physical work on reproductive ability, which have also been observed for men, suggested that differences in the natural fertility of populations, historically and today, may be explained by a direct link between food intake and fertility (Frisch 1975, 1978). This seems paradoxical and many people ask: if undernutrition decreases fertility, why are the undernourished populations of developing countries rapidly increasing? The rapid population growth currently being experienced by underdeveloped countries results from a decrease in the death rate, mainly from needed improvements in public health. The total fertility rates in developing countries are actually below the maximum human fertility rates observed in well-nourished, noncontracepting populations, as will be discussed below.

The essential finding that arose from two decades of research on these topics, is that both too little and too much fat are associated with infertility. I have hypothesized that these associations are causal, and that normally, the high percentage of body fat (26 percent to 28 percent in United States women after completion of growth) is necessary for reproduction and may influence it directly (Frisch 1984, 1985b, 1988, 1990). Both the absolute and relative amounts of fat are important, because the lean mass and the fat must be in a particular absolute range, as well as relative range, i.e., the female must have "grown up" to be big enough to successfully reproduce.

Historical Background of Research on Body Size and Fertility

Almost a century ago, Marshall and Peel (1908) reported that "over-fat" heifers, sheep, pigs and mares were sterile, and that their sterility was reversible with starvation. Experimental evidence that undernutrition also affected reproduction in animals was reported in 1922 by Evans and Bishop. They concluded that "underfeeding affects the time of maturity and ovulation history" (Evans and Bishop 1922).

Recounting "A failure and what it taught," George Corner (1946) described how, in 1929, he failed to get a response when he injected his first extracts of progesterone into rabbits, even though his collaborator, Willard Allen, succeeded with the same extract, also injected into rabbits. After months of checking and rechecking the chemistry of the extract, he found a simple explanation: when Willard Allen went to the cages to inject them, "his idea of what constitute[d] a nice rabbit led him to choose the larger ones."

Corner had a preference for the smaller ones (less than 800 gms) which were insensitive to the hormone.

The observation that reproductive dysfunction was associated with a body that was too fat, or too thin, or too small, became important in light of the elegant experiments of the late Gordon C. Kennedy. Kennedy and Mitra (1963) found that the onset of puberty was closely related to body weight, specifically, to food intake per unit of body weight, which Kennedy related to the storage of fat, a "lipostat." Kennedy hypothesized that changes in energy metabolism related to the storage of fat may be one of the signals relating body maturation to the hypothalamic control of ovarian function, establishing the onset of fertility.

Following the lead of Kennedy's research, I found that too little or too much body fat affected the reproductive ability of women. As discussed above, weight loss in the range of 10 percent to 15 percent of normal weight for height, equivalent to a loss of one third of body fat, results in the cessation of menses and ovulation (Frisch and McArthur 1974).

In addition to the disruptive effects of weight loss and athletic activity on the menstrual cycle, women who exercise moderately or who regain weight into the normal range may have menstrual cycles with a shortened luteal phase or a cycle that is anovulatory (Cumming et al. 1985; Prior 1985). These partial or total disruptions of reproductive ability are usually reversible after weight gain or reduction of physical activity.

Excessive fatness is also associated with infertility in women (Hartz et al. 1979; Pasquali et al. 1989), and weight loss restores fertility of women.

Why Fat? Caloric Cost of Reproduction

Human pregnancy and lactation both have high caloric costs: a pregnancy requires about 50,000 calories over and above normal metabolic requirements. Lactation requires an additional 1000 calories a day. In premodern times lactation was an essential part of reproduction (Frisch 1985a).

During growth the body changes in composition, as well as in size and proportions. Girls experience a continuous decline in the proportion of body water to overall body weight, because girls have a large relative increase in body fat and fat has little water (Friis-Hansen 1965). This increase in fatness is particularly rapid during the adolescent growth spurt which precedes menarche (Frisch et al. 1977).

At the completion of growth, between the ages of 16 and 18, the body of a well-nourished woman contains about 26 percent to 28 percent fat and about 52 percent water, whereas the body of a man at completion of growth contains about 14 percent fat and 61 percent water (Moore et al. 1963;

FIGURE I

Changes in Body Water As Percentage of Body Weight Throughout the Life Span, and Corresponding Changes in the Percentage of Body Fat.

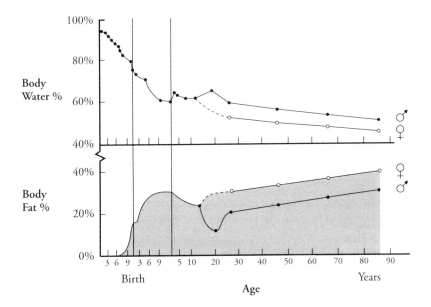

Adapted from Friis-Hansen, B. (1965). "Hydrometry of growth and aging." In *Human Body Composition. vol. VII.* (J. Brozek, ed.), 191–209. Pergamon Press, Oxford. Reprinted with permission.)

TABLE I

Total Body Water as Percent of Body Weight, and Index of Fatness: Comparison of an 18-Year-Old Girl and a 15-Year-Old Boy of the Same Height and Weight

Variable	Girl	Boy
Height (cm)	165.0	165.0
Weight (kg)	57.0	57.0
Total body water (liter)	29.5	36.0
Lean body weight (kg)[a]	41.0	50.0
Fat (kg)	16.0	7.0
Fat/body weight (%)[b]	28.0	12.0
Total body water/body weight (%)	51.8	63

a   Lean body weight = total body water/0.72
b   Fat/body weight % = 100 − [(total body water/body wt %)/0.72]

Edelman et al. 1952) (See Figure 1). A young girl and boy of the same height and weight, as shown in Table 1, differ markedly in their percentages of body water and fat (Frisch 1981). Females store an average of 16 kg of fat, which is equivalent to 144,000 calories. The main function of this fat may be to provide energy for a pregnancy and about 3 months of lactation. In prehistoric times, when the food supply was scarce or fluctuated seasonally, stored fat was necessary for successful reproduction (Frisch 1984, 1985b).

The survival of an infant is correlated with its birth weight, and birth weight is correlated with the pre-pregnancy weight of the infant's mother, and independently, her weight gain during pregnancy (Eastman and Jackson 1968). One can therefore regard the turning off of the hypothalamic signal, which is necessary to activate the ovary, as adaptive, because a woman who weighed too little would have a child who weighed too little, and who would thus have a low chance of survival. The finding that the pituitary-ovarian axis is intact even though the hypothalamic signal is lacking supports this view (Vigersky et al. 1977). Weight gain restores the hypothalamic regular signal of releasing hormone and the pituitary-ovary axis responds with ovulation and regular menses, thus restoring fertility. Essentially, weight loss and excessive leanness result in a return to a prepubertal pattern of endocrine secretion, normally found with prepubertal, lean body composition. The normal adult pattern returns with the return of adult female body composition.

The findings of Van der Spuy et al. (1988) that women in whom ovulation had been induced had a higher risk of having babies who were small for date, and that this risk was greatest (54 percent) in those women who were underweight also supports the view that hypothalamic infertility is adaptive. Van der Spuy et al concluded that the most suitable treatment for infertility caused by weight related amenorrhea (absence of ovulation and menstrual cycles) is controlling the diet, not inducing ovulation.

Over a century ago, Dr. J. Matthew Duncan (1871) summed up these relationships for the Royal College of Physicians: "If a seriously undernourished woman could get pregnant the chance of her giving birth to a viable infant, or herself surviving the pregnancy is infinitesimally small."

How Adipose Tissue may Regulate Female Reproduction

There are at least four mechanisms by which adipose tissue (body fat) may directly affect ovulation and the menstrual cycle, and hence fertility:

1. Adipose tissue is a significant extragonadal source (i.e., in addition to the estrogen produced by the ovary) of estrogen, the most important female hormone (Siiteri 1981; Siiteri and MacDonald 1973). Conversion of

androgen (a male hormone) to estrogen takes place in adipose tissue of the breast and abdomen (Nimrod and Ryan 1975), the omentum (Perel and Killinger 1979), and the fatty marrow of the long bones (Frisch et al. 1980a). This conversion accounts for roughly a third of the circulating estrogen of premenopausal women, and is the main source of estrogen in postmenopausal women (Siiteri and MacDonald 1973). Men also convert androgen into estrogen in body fat.

2. Body weight influences the direction of estrogen metabolism to more potent or less potent forms (Fishman et al. 1975). Very thin women have an increase in the 2-hydroxylated form of estrogen, a nonpotent estrogen, which has a low activity and little affinity for the estrogen receptor. Lean women athletes also have an increase in the nonpotent form of estrogen (Snow et al. 1989, Frisch et al. 1992). In contrast, obese women produce less of the nonpotent form and have a relative increase in the 16-hydroxylated form, which has potent estrogenic activity (Schneider et al. 1983).

3. Obese women (Siiteri 1981) and young girls who are relatively fatter than average (Apter et al. 1984) have a diminished capacity for producing estrogen to bind to serum sex-hormone-binding globulin (SHBG); this results in an elevated percentage of free plasma estrogen. Since SHBG regulates the availability of estradiol to the brain and other target tissues, the changes in the proportion of body fat to leanness may influence reproductive performance through the intermediate effects of SHBG.

4. Adipose tissue of obese women stores steroid hormones (Kaku 1969).

Changes in relative fatness might also affect reproductive ability indirectly through disturbance of the regulation of body temperature and energy balance by the hypothalmus. Very lean women, both anorectic and nonanorectic, display abnormalities of temperature regulation, in addition to delayed response or lack of response to exogenous gonadotropin-releasing hormone (GnRH) (Vigersky et al 1977).

Hypothalamic Dysfunction, Gonadotropin Secretion, and Weight Loss

It is now known that the amenorrhea of underweight and excessively lean women, including athletes, is due to hypothalamic dysfunction. The pituitary-ovarian axis in excessively lean women is apparently intact, and functions when exogenous gonadotropin releasing hormone (GnRH) is given (Nillius 1983) [See Hypothalamus in Glossary].

Girls and women with this type of hypothalamic amenorrhea experience both quantitative and qualitative changes in the secretion of gonadotropins, luteinizing hormone (LH), and follicle-stimulating hormone (FSH): (1) LH and FSH are low; estradiol levels are also low; (2) the secretion of LH and the response to GnRH are reduced in direct correlation with the amount of weight loss (Vigersky et al. 1977); (3) underweight patients respond to exogenous GnRH with a pattern of secretion similar to that of prepubertal children; the FSH response is greater than the LH response. The return of LH responsiveness is correlated with weight gain (Warren et al. 1975); (4) the maturity of the 24-hour LH secretory pattern and body weight are related; weight loss results in an age-inappropriate secretory pattern resembling that of prepubertal or early pubertal children. Weight gain restores the post-menarcheal secretory pattern (Boyar et al. 1974).

The Physiological Basis of Reproductive Ability: Weight at Menarche

The idea that relative fatness is important for female reproductive ability followed from the findings that the events of the adolescent growth spurt, particularly menarche in girls, were closely related to an average critical body weight (Frisch and Revelle 1971b). The physiological basis of reproductive ability is also similar to the weight-dependencies of estrus (the start of reproductive ability) of many mammals (Kennedy and Mitra 1963).

The mean weight at menarche for a sample of middle class U.S. girls was 47 kg, at the mean height of 158.5 cm at the mean age of 12.9 years. This mean age included girls from Denver, who had a slightly later age of menarche than the sea-level populations because of the slowing effects of altitude on rate of weight growth. The average age of menarche of U.S. girls is now 12.6–12.8 years (Frisch and Revelle 1971b).

The Secular (Long-Term) Trend Toward an Earlier Age of Menarche

Even before analyzing the meaning of the average critical weight for an individual girl, the idea that menarche is associated with a critical weight for a population explained many observations associated with early or late menarche. Observations of earlier menarche are associated with girls attaining the critical weight more quickly. The most important example is the secular (long-term) trend to an earlier menarche of about three or four months per decade in Europe in the past 100 years (See Figure 2). Our

FIGURE 2

Mean or Median Age of Menarche As a Function of Calendar Year From 1790 to 1980.

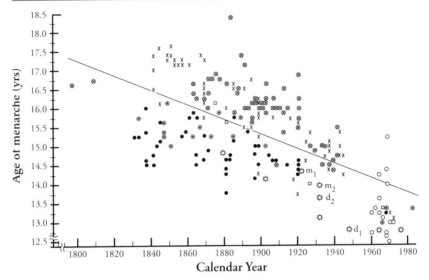

The symbols refer to England (⊙), France (●), Germany (⊗), Holland (◻), Scandinavia (x) (Denmark, Finland, Norway, and Sweden); Belgium, Czechoslovakia, Hungary, Italy, Poland (rural), Romania, (urban and rural), Russia (15.2 years at an altitude of 2500m and 14.4 years at 700m), Spain and Switzerland (all labeled O); and the United States (✿), data not included in the regression line.) Twenty-seven points for Europe were identical and do not appear on the graph. The regression line, of course, cannot be extended indefinitely. The age of menarche has already leveled off in some European countries, as it has in the United States (see text). [Reprinted from Wyshak and Frisch (1982.)]

explanation is that children nowadays become bigger sooner; therefore, girls on the average reach 46–47 kg, the average weight at menarche of the U.S. and many European populations, more quickly. The secular trend should end when the rate of weight growth of children of successive cohorts remains the same because of the attainment of maximum nutrition and child care, as has now happened in the United States.

Conversely, a late menarche is associated with body weight growth that is slower prenatally, postnatally, or both, so that the average critical weight is reached at a later age. Undernutrition delays menarche; twins have later menarche than do singletons of the same population; and high altitude also delays menarche. Diseases, such as cystic fibrosis, juvenile onset diabetes, Crohn's disease and sickle cell anemia have also been shown to delay menarche (Frisch and Revelle 1971b).

Components of Weight at Menarche

Individual girls have menarche at varied weights and heights. To predict menarche for an individual girl, we investigated body composition at menarche, because components of body weight, total body water (TBW) and lean body weight (LBWt) are more closely correlated with metabolic rate than is body weight (BWt). Lean body weight (LBWt) is calculated from total body water (TBW) divided by a constant, 0.72 (Moore et al. 1963).

Metabolic rate was considered an important clue since the late G. C. Kennedy hypothesized a food intake-lipostat-metabolic signal related to stored fat to explain his findings on weight and puberty in rats (Kennedy and Mitra 1963).

Roger Revelle and I made an extensive study of changes in body composition (total water as percent of body weight, lean body mass and fat mass) in a sample of 181 girls for whom data were available from three longitudinal growth studies that followed them from birth to age 18. We found that the greatest change in estimated body composition of both early- and late-maturing girls during the adolescent growth spurt was a large increase in body fat, from about 5–11 kg, a 120 percent increase, compared to a 44 percent increase in lean body weight. Body fat has little water (about 10 percent) compared with muscle and viscera (about 80 percent). Therefore, as the body increases in fatness, the percentage of total water in the body decreases. We found that the ratio of lean body weight (LBWt) to fat fell from 5:1 at initiation of the spurt to 3:1 at menarche (Frisch et al. 1973).

Relative Fatness as a Determinant of Minimal Weights for Menstrual Cycles

I also found that I could predict minimum weights for the onset and maintenance of menstrual cycles from the indicator for fatness, total water as a percentage of body weight (Frisch and McArthur 1974). Percentiles of total body water/body weight percent, (TBW/BWt%) were made at menarche for 181 girls and for the same girls again at age 18, the age at which body composition was stabilized.

Patients with amenorrhea (absence of cycles) due to weight loss, other possible causes having been excluded, were studied in relation to the weights indicated by the diagonal percentile lines of total water/body weight percent in Figure 3. We found that 56.1 percent of total water/body weight, the tenth percentile at age 18 years, which is equivalent to about 22 percent fat of body weight, indicated a minimal weight for height necessary for the restoration and maintenance of menstrual cycles. For example, a 20-year-old woman whose height is 165 cm (65 inches) should weigh at least 49 kg (108 lbs) before menstrual cycles would be expected to resume (See Figure 3).

The weights at which menstrual cycles ceased or resumed in post-menarcheal patients aged 16 years and older were about 10 percent heavier than the minimal weights for the same height observed at menarche (Frisch and McArthur 1974). This is because both early- and late-maturing girls gain an average of 4.5 kg of fat from menarche to age 18. At age 18, mean fat is 16.0 kg, 28 percent of the mean body weight of 57 kg. Reflecting this increase in fatness, the total water/body weight percent decreases from 55 percent at menarche (12.9 ± 0.1 year in our sample) to 52 percent at age 18 (Frisch 1976).

Because girls are less fat at menarche than when they achieve stable reproductive ability, the minimal weight for onset of menstrual cycles in cases of primary amenorrhea (delayed menarche) due to undernutrition or exercise is indicated by the 10th percentile of fractional body water at menarche, 59.8 percent, which is equivalent to about 17 percent of body weight as fat. For example, a 15-year-old girl whose completed height is 165 cm (65 inches) should weigh at least 44 kg (97 lb) before menstrual cycles can be expected to begin. (See Figure 3, "3 & 4"). These minimum weights would also be used for girls who become amenorrheic as a result of weight loss shortly after menarche, as is often found in cases of anorexia nervosa in adolescent girls. The absolute and relative increase in fatness from menarche to ages 16–18 years coincides with the period of adolescent subfecundity. During this time there is still rapid growth of the uterus, the ovaries, and the oviducts (Scammon 1930).

These minimal weights for heights are now used clinically in the evaluation and treatment of patients with primary or secondary amenorrhea due to weight loss or excessive leanness (Nillius 1983).

The prediction of the minimum weights for height is from the percentage of Total Body Water divided by body weight (TBW/BWt%)—not the percentage of fat divided by the body weight—thus indicating that the size of the lean mass is important in relation to the amount of fat (i.e., the prediction is based on a lean mass:fat ratio). This ratio is about 3:1 at menarche and 2.5:1 at the completion of growth at age 18 years. No prediction can as yet be made above the threshold weight for a particular height.

Other factors, such as emotional stress, affect the maintenance or onset of menstrual cycles. (Reichlin 1982). Therefore, menstrual cycles may cease without weight loss and may not resume in some subjects even though the minimum weight for height has been achieved. Also, these minimal weight standards apply thus far only to Caucasian women in the United States and Europe, since different races have different critical weights, and it is not yet

FIGURE 3

Reproduction and Fertility

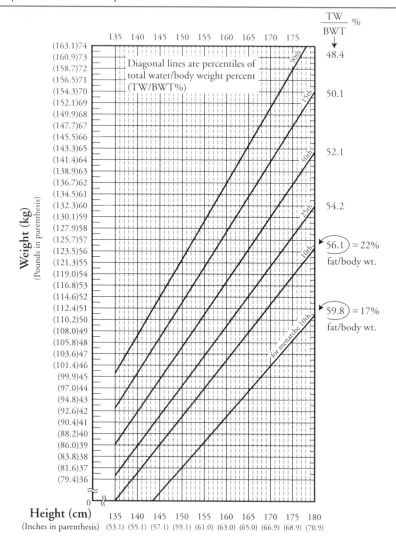

The minimal weight necessary for a particular height for restoration of menstrual cycles is indicated on the weight scale by the 10th percentile diagonal line of total water/body weight percent (TW/BWt %), 56.1, as it crosses the vertical height line. For example, a 20-year-old woman whose height is 165 m (65 in) should weigh at least 49 kg (108 lb) before menstrual cycles would be expected to resume. The minimal weight necessary for a particular height for onset of menstrual cycles is indicated on the weight scale by the 10th percentile diagonal line of TW/BWt %, 59.8, as it crosses the vertical height lines. Height growth of girls must be completed, or approaching completion. For example, a 15-year-old whose completed height is 165 m (65 in) should weigh at least 43.6 kg (96 lb) before menstrual cycles can be expected to start.
(Adapted from Frisch and McArthur (1974) with permission from Science.)

known whether the different critical weights represent the same critical body composition of fatness. (Frisch and McArthur 1974).

Some amenorrheic athletes, such as shotputters, oarswomen, and some swimmers, are not lightweight because they are very muscular, and muscles are heavy (80 percent water) compared with fat (about 10 percent water). The cause of their amenorrhea is, nevertheless, most probably the increased lean mass and reduced fat content of their bodies; gaining body fat or ceasing exercise usually restores menstrual cycles. In relation to athletic amenorrhea, it is important to note that body composition may change without any change in body weight. A woman may increase muscle mass by increasing training and at the same time lose fat, without a perceptible change in body weight (Frisch et al 1993).

Physical Exercise, Delayed Menarche and Amenorrhea

We found that ballet dancers (Frisch et al 1980b) and young prepubertal athletes experience delayed menarche. Other researchers confirmed these findings (Frisch et al 1981), and they became widely known in the last decade.

The mean age at menarche in a sample of 38 college swimmers and runners was 13.9 ± 0.3 years, significantly later than that of the general population, 12.8 ± 0.05 years, in accord with other reports. However, the mean menarcheal age of the 18 athletes whose training began before menarche was 15.1 ± 0.5 years, significantly later than the mean menarcheal age of the 20 athletes whose training began after menarche, which was 12.8 ± 0.2 years. The latter mean age was similar to that of the college controls, 12.8 ± 0.4 years, and the general population. Therefore, training, not pre-selection, is the delaying factor. Each year of premenarcheal training delayed menarche by 5 months (0.4 yr). The training also disrupted the regularity of the menstrual cycles in both pre- and postmenarcheal trained athletes. Athletes with irregular cycles or amenorrhea had hormonal levels confirming lack of ovulation, hence infertility.

This suggests that one constructive way to reduce the incidence of teenage pregnancy would be to have girls maintain regular moderate exercise beginning at age 8 or 9. Such a program might also reduce the risk of serious diseases of women in later life, such as breast cancer, cancers of the reproductive system and diabetes, as will be discussed below.

Long Term, Regular Exercise Lowers the Risk of Sex Hormone Sensitive Cancers

Our findings of irregular menstrual cycles, or absence of cycles among young women athletes (Frisch et al. 1981) raised the question: "Are there differences

FIGURE 4

Prevalence Rate (Lifetime Occurrence Rate) of Cancers of the Reproductive System for Athletes and Nonathletes by Age Group.

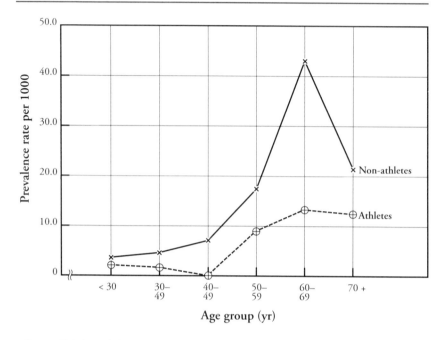

Source: Frisch et al. (1985). Br. J. Cancer, 52, 885.)

in the long-term reproductive health of moderately trained athletes compared to non-athletes?"

A study of 5398 college alumnae aged 20 to 80 years, 2622 of whom were former athletes and 2776 of whom were nonathletes, showed that the former athletes had a significantly lower lifetime occurrence (prevalence) of breast cancer and cancers of the reproductive system than the nonathletes. Over 82.0 percent of the former college athletes began their training in high school or earlier, compared to 25 percent of the nonathletes. The analysis controlled for potential confounding factors such as age, age at menarche, age at first birth, smoking, cancer family history, etc (Frisch et al. 1985). The sedentary women had 2.5 times the risk of cancers of the reproductive system as the former college athletes. The risk of breast cancer among the sedentary women was about twice that of the former college athletes. (See Figure 4). The former athletes were leaner in every age group than the nonathletes.

The most likely explanation for this lower risk is that the former athletes had lower levels of estrogen for many years, particularly in adolescence, because they were leaner, and more of the estrogen was metabolized to the nonpotent form of estrogen (Frisch et al. 1985; Hershcopf and Bradlow 1987). Also, the former athletes may have consumed diets lower in fat and saturated fat (Frisch et al. 1981). Such diets shift the pattern of estrogen metabolism toward the less active form of estrogens (Longcope et al. 1987).

The former college athletes also had a lower lifetime occurrence of benign tumors of the breast and reproductive system (Wyshak et al. 1986), and a lower prevalence of diabetes, particularly after age 40, than the nonathletes (Frisch et al. 1986). Leaner women are at lower risk for diabetes.

These data indicate that long term exercise, which was not at Olympic or marathon level but moderate and regular, reduces the risk of sex hormone sensitive cancers, and the risk of diabetes for women later in life. Recent data which show that moderate exercise also reduces the risk of nonreproductive system cancers (Frisch et al. 1989; Blair et al. 1989), suggest that other protective factors, such as changes in the immune system may also be involved.

Current Research: Magnetic Resonance Imaging (MRI) of Body Fat

The association of the disruption of reproductive ability with changes in body composition has been controversial because methods of quantifying body fatness were necessarily indirect, deriving from the measurement of total body water.

Magnetic Resonance Imaging (MRI) is a noninvasive method of quantifying both subcutaneous and internal fat. We have recently shown that the athletes in our study had 30 percent to 40 percent less fat than the controls although the mean body weight of women athletes in our samples did not differ from that of sedentary controls and was even heavier in some studies (Gerard et al. 1991, Frisch et al. 1993). Also, by quantifying the body fat by MRI, we showed the extent of metabolism to the nonpotent estrogen, 2-hydroxyestrone, to be significantly ($P<0.003$) inversely related to the amount of body fat, thus indicating a pathway from the relative amount of body fat to the hypothalamic control of reproduction (Frisch et al. 1993).

Double-Muscled Cattle and Other Animal Data

That loss of fat and extreme leanness, not weight loss per se, is the important factor in infertility, can be deduced from the relative infertility of the breed of double-muscled, Charolais cattle. As described by Vissac, et al. (1974), female cattle with a greater muscular mass have evident physiological troubles with their puberty, their fertility, and their sexuality in general.

Rats fed a high-fat (HF) diet, the fat being substituted isocalorically for carbohydrate, had estrus significantly earlier, p<0.0001, than did rats fed a low-fat (LF) diet (Frisch et al. 1975). Direct carcass analysis data showed that the HF and LF diet rats had similar body compositions at estrus, although the HF rats had estrus at a lighter body weight than the LF diet rats (Frisch et al. 1977).

Link Between Food Intake and Ovulation: "Flushing"

"Flushing" is the increase in the rate of twinning in sheep resulting from short-term (e.g. one week) high-caloric feeding of the ewe before mating to the ram (Coop 1966). The well-nourished human female fortunately does not normally have multiple ovulations in response to high caloric intake, such as a large steak dinner. However, there is evidence that some residual flushing effect does occur in human beings. The rate of human dyzygotic twinning, which results from two independent ovulations, but not monozygotic twinning (one egg dividing into two parts), fell during wartime restrictions of nutrition in Holland; the dyzygotic rate returned to normal after the return of a normal food supply (Bulmer 1970).

Effects of Nutrition and Exercise on Male Reproductive Function

Undernutrition delays the onset of sexual maturation in boys (Tanner 1962), just as it delays a girl's age of menarche. Undernutrition and weight loss in men also affects their reproductive ability. The sequence of effects, however, is different from that of females. In men, loss of libido is the first effect of a decrease in caloric intake and subsequent weight loss. Continued caloric reduction and weight loss result in a loss in prostrate fluid, and decrease of sperm mobility and longevity, in that order. Sperm production ceases when weight loss is in the range of 25% of normal body weight. Caloric increase results in a restoration of function in the reverse order of loss (Keys et al. 1950).

Male marathon runners experience a decrease in hypothalamic gonadotropin-releasing hormone secretion (GnRH) similar to that of women runners (MacConnie et al. 1986). Also reported are changes in serum testosterone levels with weight loss in wrestlers (Strauss 1985); a reduction in serum testosterone and prolactin levels in male distance runners (Wheeler 1984), and changes in reproductive function and development in relation to physical activity (Wall and Cummings 1985). Hypothalamic dysfunction and hormonal changes in the male are apparently associated with more intense athletic training compared to the levels causing hypothalamic dysfunctions in female athletes.

Nutrition, Physical Work, and Natural Fertility: Human Reproduction Reconsidered?

The effects of physical work and nutrition on reproductive ability suggested that differences in the fertility of populations, historically and today, may be explained by a direct pathway from food intake to fertility, in addition to the classic Malthusian pathway through mortality. Charles Darwin (1868) described this common sense relationship, observing that (1) domestic animals, that have a regular, plentiful food supply without working for it, are more fertile than wild animals; (2) " hard living retards the period at which animals conceive;" (3) the amount of food affects the fertility of the same individual; and (4) it is difficult to fatten a cow that is lactating (Darwin 1859). All of Darwin's dicta apply to human beings.

The Paradox of Rapid Population Growth in Undernourished Populations

In many historical populations with slow population growth, poor couples living together to the end of their reproductive lives had only six to seven live births. Most poor couples in many developing countries today also have only six to seven live births during their reproductive lifespan. But six children per couple today in developing countries results in a rapid rate of population growth because of decreased mortality rates . The difference between the birth rate per 1000 and the death rate per 1000, (the growth rate per 1000), is now as high as 2 percent, 3 percent and 4 percent in many developing countries. Populations growing at these rates double in 35, 23 and 18 years, respectively (references in Frisch 1978).

Variation in the natural fertility of populations is well-established (Campbell and Wood 1988). The total fertility rate (the number of live births that a married couple produce during their entire reproductive lives together) can be as low as the four observed among the nomadic !Kung people of the Kalahari desert. A total fertility rate of six or seven births, typical of many developing countries today, is far below the human maximum of 11 and 12 children found, on average, among well-nourished, noncontracepting Hutterites (See Figure 5). The usual explanation of the lower-than-maximum fertility observed in both historical and contemporary societies is the use of "folk" methods of contraception, abortion or venereal disease, in combination with social customs that can affect fertility, such as late age of marriage or a taboo on intercourse during lactation. Because food intake and energy outputs can directly affect fecundity, undernutrition and hard physical work are an alternative explanation of the observed submaximum fertility (See Figure 6).

FIGURE 5

The Curve of Reproductive Ability (Maximum fertility rate = 100)

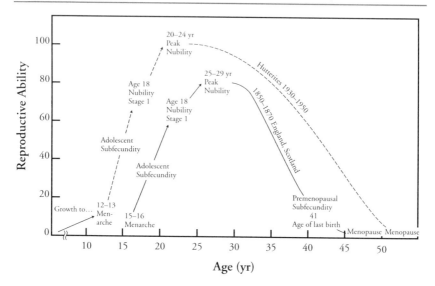

The mid-nineteenth century curve of female reproductive ability (variation of the rate of childbearing with age) compared to that of the well-nourished, non-contracepting modern Hutterites. The Hutterite fertility curve results in an average of 10 to 12 children; the 1850–1870 fertility curve in about six to eight children. (Reprinted from Frisch (1978) with permission from Science).

Slow growth to maturity of women and men, caused by undernutrition, hard physical work and disease, is correlated with a shorter, less efficient reproductive span, than that of well-nourished populations (See Figure 5).

Historical data from developed countries and contemporary data from developing countries show that compared to well-nourished populations, females who grow relatively slowly to maturity, completing height growth at ages 20–21 years (instead of 16–18 years as in well-nourished contemporary populations), differ from well-nourished females in each event of the reproductive span: Menarche is later (e.g., 15.0-16.0 years, compared with 12.8 years); adolescent sterility is longer; the age of peak nubility is later; the levels of specific fertility are lower; pregnancy wastage is higher; the duration of lactational amenorrhea is longer; the birth interval is therefore longer; and the age of menopause is earlier, preceded by a more rapid period of perimenopausal decline. The average age of menopause in the United States today is 52 years, compared to about 45 to 47 years historically (Frisch 1978).

FIGURE 6

Biological and Social Factors Influencing the Mean Number of Live Births per Couple (Total Fertility Rate)

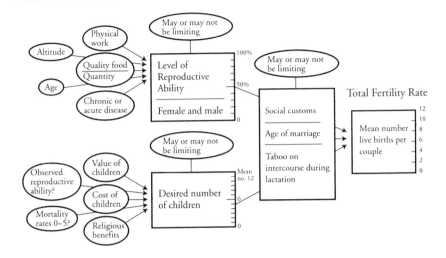

The mean number of live births per couple (Total Fertility Rate) may be influenced by the level of reproductive ability and the desired number of children. Each of these factors is influenced, in turn, by environmental factors such as nutrition, intrinsic factors such as age, and social factors such as the value of children. Each of the major factors may be limiting alone or together, or in interaction with social customs such as age of marriage. Social customs may or may not be limiting. Well-nourished couples may have a level of reproductive ability of 100 percent and yet have only two children because that is their desired number, and efficient contraception is used. Poorly nourished couples may have a total fertility rate of about five, as is found among Bush people of the Kalahari desert; however, the desired number is apparently greater than five. (The Bush women say, "God is stingy with children.") The main limiting factor for the Bush women may be a reduced level of reproductive ability due to relative undernutrition and high levels of physical activity. (From Frisch, R.E. Population, Nutrition and Fecundity. In N. Eberstadt, ed. Fertility Decline in the Less Developed Countries, Praeger, 1981, New York.)

Thus the slower, submaximum growth of women to maturity is subsequently associated with a shorter, less efficient reproductive span. The faster the growth of the females and males, the earlier and more efficient the reproductive ability. Natural fertility therefore would be expected to rise as higher socioeconomic levels are attained. For example, data show an increase in natural fertility in Taiwan, which accompanied recent improvements in health and nutrition (Jejeebhoy 1983). Therefore, the need for family planning programs, when there is the hoped for progress in standards of living, may be much greater during the transitional period to a lower family size.

Necessity for Efficient Methods of Contraception

One of the most important differences observed between the reproductive pattern of well-nourished women and that of less well-nourished women, is that well-nourished lactating women who nurse their children on demand, and without supplementation, may be at risk of becoming pregnant again within three to four months after the birth of a child (Fitzgerald 1992), whereas in the past, and at present in poorly nourished populations, women who are lactating do not usually resume ovulatory cycles before a year or a year and a half after giving birth. Closely spaced pregnancies are dangerous to the health of the nursing infant because the milk supply decreases when there is a new pregnancy. They are also dangerous to the mother and the fetus because lactation is an energy drain and the mother will not have had sufficient time to restore the depleted tissues. Thus a too-short birth interval can result in infants of low birth weight, increasing the risk of postnatal mortality and diseases (Frisch 1985a; Bracher 1992).

As nutrition and general health improve, women are fecund longer than in the past, since menopause is later. Pregnancies late in life are more dangerous to both the mother and her infant, particularly if there have been many previous pregnancies.

Thus, unavailability of modern, reliable methods of contraception or religious bans on the use of modern methods of contraception can endanger the health of both the mother and her infant. For example, the rhythm method of birth spacing is not suitable or protective during lactation because a woman may have irregular menstrual cycles when she resumes menstrual cycles while she is lactating, or for some months after completion of lactation.

The rhythm method is also not suitable or protective during the years of the menopausal period, which usually begins at age 40, because the length of the menstrual cycle begins to vary as cycles become irregular during the premenopausal period. Some proponents of the rhythm method now recommend the use of evaluation of the thickness or thinness of the cervical mucus (Billings method) to make the rhythm method more precise. The woman is then supposed to inform her husband when they cannot have intercourse if they desire to avoid pregnancy. Since it is contrary to the mores of most women in developing countries to evaluate their cervical mucus each day, and even more contrary to their mores to discuss such matters with their husbands, the unavailability of alternatives to the rhythm method can be considered a danger to the health of mothers and infants in developing countries (Frisch, Letters to the Editor, NY Times 1987). The rise in the incidence of AIDS also adds to the dangers of the rhythm method.

Conclusion

Roger Revelle, I think, would be pleased at the number of spinoffs relevant to public health and reproductive health from our initial findings. Among them, the nomograms predicting minimum weights for height for menarche and the restoration of menstrual cycles and ovulation are reprinted in many obstetrical textbooks for the onset of cycles and to help women restore fertility. The first study of the long-term health of American college women resulted in the landmark finding that moderate, regular exercise reduces the risk of breast cancers (recently confirmed by Bernstein et al. 1994), and cancers of the reproductive system in women. Former athletes also had a reduced risk of noninsulin dependent diabetes. The first direct quantification of overall body fat with magnetic resonance imaging of athletes vs. controls showed that body weight and the body mass index can be a misleading index of body composition: athletes can have 30–40% less body fat and still weigh the same as sedentary controls. The metabolism of estrogen to a nonpotent form was shown to be inversely related to total fat/total volume percent. This direct role of adipose tissue in relation to the control of estrogen metabolism is being explored by Bradlow et al. (1992). Also, the recent cloning of the ob (obese) gene in mice and the human homologue, showed that adipose tissue secretes a protein, leptin, which regulates body weight by controlling food intake and energy expenditure (Halaas et al 1995). These new data strongly support the concept of a lipostat, with a feedback to the hypothalmus which controls both food-intake and reproduction.

References

Apter, D., N. J. Bolton, G. L. Hammond, et al. "Serum Sex Hormone Binding Globulin During Puberty in Girls and in Different Types of Adolescent Menstrual Cycles." *ACTA Endocrinologica* 107 (1984):413–19.

Bernstein, L., B. E. Henderson, R. Hanish, J. Sullivan-Halley and R. H. Ross. "Physical Exercise and Reduced Risk of Breast Cancer in Young Women." *Journal of the National Cancer Institute* 86(1994):1403–08.

Blair, S. N., H. W. Kohl III, R. S. Paffenbarger, et al. "Physical Fitness and All-cause Mortality." *J. Am. Med. Assoc.* 262 (1989):2395–2401.

Boyar, R.M., J. Katz, J.W. Finkelstein, et al. "Anorexia Nervosa: Immaturity of the 24-hour Luteinizing Hormone Secretory Pattern." *N.Engl. J. Med.* 291 (1974):861–65.

Bracher, M. "Breastfeeding, Lactational Infecundity, Contraception, and the Spacing of Births: Implications of the Bellagio Consensus Statement." *Health Transition Review* 2 (1992):19–42.

Bradlow, H. L., N. T. Telang and A. Sutro. "Modulation of Estrogen Metabolism in Estrogen Responsive MCF-7 Cell by Adipocyte Conditioned Media." *Proceedings of American Association of Cancer Research*, Program, Abstract, Naples, Florida, 1992.

Bulmer, M. G. *The Biology of Twinning in Man.* Oxford: Oxford University Press, 1970.

Campbell, K. L., and J. W. Wood. "Fertility in Traditional Societies." In *Natural Fertility*, P. Diggory, M. Potts and S. Teper, eds. London, MacMillan Press, 1988.

Cohn, S. H., A. N. Vaswani, S. Yasumura, et al. "Improved Models for Determination of Body Fat by In Vivo Neutron Activation." *Am. J. Clin. Nutr.* 40 (1984):255–59.

Coop, I. E. "Effect of Flushing on Reproductive Performance in Ewes." *J. Agr. Science.* 67 (1966):305–23.

Corner, G. W. *The Hormones in Human Reproduction.* Princeton, Princeton University Press, 1946:118–20.

Cummings, D. C., M. M. Vickovic, S. R. Wall, et al. "Defects in Pulsatile LH Release in Normally Menstruating Runners." *J. Clin. Endocrin. Metab.* 6 (1985):810–12.

Darwin, C. *The Origin of the Species.* Cambridge, MA: Harvard University Press, 1975, 147.

———. *The Variation of Animals and Plants Under Domestication.* Appleton, New York, (1894) edition 2, Vol. 2:89–90.

Duncan, J. M. *Fecundity, Fertility, Sterility and Allied Topics*, ed 2. Edinburgh: Adam and Charles Black 1871.

Eastman, N. J., and E. Jackson. "Weight Relationships in Pregnancy. I. The Bearing of Maternal Weight Gain and Pre-Pregnancy Weight on Birth Weight in Full Term Pregnancies." *Obstetrical and Gynecology Survey.* 23 (1968):1003–1025.

Edelman, I. S., H. B. Haley, P. R. Scholerb, et al. "Further Observations on Total Body Water. I: Normal Values Throughout the Life Span." *Surg. Gynecol. Obstet.* 95 (1952):1–12.

Evans, H.M., and K.S. Bishop. "On the Relations Between Fertility and Nutrition. II: The Ovulation Rhythm in the Rat on Inadequate Nutritional Regimes." *J. Metab. Res.* 1 (1922):335–56.

Fishman, J., R. M. Boyar, and L. Hellman. "Influence of Body Weight on Estradiol Metabolism in Young Women." *J. Clin. Endocrinol. Metab.* 41 (1975):989–91.

Fitzgerald, M. H. "Is Lactation Nature's Contraceptive? Data from Samoa." *Social Biology* 39 (1992):55–64.

Friis-Hansen, B. "Hydrometry of Growth and Aging." In *Human Body Composition*, vol VII, Symposia of the Society for the Study of Human Biology, J. Brozek ed. Oxford, Pergamon, 1965:191–209.

Frisch, R. E. "Weight at Menarche: Similarity for Well-nourished and Under-nourished Girls at Differing Ages and Evidence for Historical Constancy." *Pediatrics* 50 (1972):445–50.

_____ . "Demographic Implications of the Biological Determinants of Female Fecundity." *Social Biology* 22 (1975):17–22.

_____ . "Fatness of Girls From Menarche to Age 18 Years with a Nomogram." *Human Biology* 48 (1976):353–359.

_____ . "Population, Food Intake and Fertility." *Science* 199 (1978):22–30.

_____ . "What's Below the Surface?" *N. Engl. J. Med.* 305 (1981):1019–20.

_____ . "Body Fat, Puberty and Fertility." *Biol. Rev.* (Gr. Britain) 59 (1984):161–88.

_____ . "Maternal Nutrition and Lactational Amenorrhea: Perceiving the Metabolic Costs." In *Maternal Nutrition and Lactational Infertility,* J. Dobbing, ed. New York: Raven Press, (1985a):65–78.

_____ . "Fatness, Menarche and Female Fertility." *Persp. Biology and Med.* 8 (1985b):611–33.

_____ . "Times When the Rhythm Method is Unreliable." *NY Times*, Letters to the Editor, Oct. 14, 1987.

_____ . "Birth Control That Few Are Able to Practice." *NY Times*, Letters to the Editor, Nov. 10, 1987.

_____ . "Fatness and Fertility." *Scientific American* 258 (1988):88–95.

Frisch, R. E., ed. *Adipose Tissue and Reproduction.* Karger:Basel, 1990.

Frisch, R. E., and R. Revelle. "Variation in Body Weights and the Age of the Adolescent Growth Spurt Among Latin American and Asian Populations in Relation to Calorie Supplies." *Human Biology* 41 (1969a):536–59.

Frisch, R. E., and R. Revelle. "The Height and Weight of Adolescent Boys and Girls at the Time of Peak Velocity of Growth in Height and Weight: Longitudinal Data." *Human Biology* 41 (1969b):536–59.

Frisch, R. E., and R. Revelle. "Height and Weight at Menarche and a Hypothesis of Critical Body Weights and Adolescent Events." *Science* 169 (1969c):397–99.

Frisch, R. E., and R. Revelle. "The Height and Weight of Girls and Boys at the Time of the Initiation of the Adolescent Growth Spurt in Height and Weight and the Relationship to Menarche." *Human Biology* 43 (1971a): 140–65.

Frisch, R. E., and R. Revelle. "Height and Weight at Menarche and a Hypothesis of Menarche." *Archiv Dis Childh* 46 (1971b):695–701.

Frisch, R. E., R. Revelle, and S. Cook. "Components of Weight at Menarche and the Initiation of the Adolescent Growth Spurt in Girls: Estimated Total Water, Lean Body Weight and Fat." *Human Biology* 45 (1973): 469–74.

Frisch, R. E., and J. W. McArthur. "Menstrual Cycles: Fatness as a Determinant of Minimum Weight for Height Necessary for their Maintenance or Onset." *Science* 185 (1974):949–51.

Frisch, R. E., D. M. Hegsted, and K. Yoshinaga. "Body Weight and Food Intake at Early Estrus of Rats on a High Fat Diet." *Proc. Natl. Acad. Sci.* 72 (1975):4172–76.

Frisch, R. E., D. M. Hegsted, and K. Yoshinaga. "Carcass Components at First Estrus of Rats on High Fat and Low Fat Diets: Body Water, Protein, and Fat." *Proc. Natl. Acad. Sci.* 74 (1977):379.

Frisch, R. E., J. A. Canick, and D. Tulchinsky. "Human Fatty Marrow Aromatizes Androgen to Estrogen." *J. Clin. Endocrinol. Metab.* 51 (1980a):394–96.

Frisch, R. E., G. Wyshak, and L. Vincent. "Delayed Menarche and Amenorrhea of Ballet Dancers." *N. Engl. J. Med.* 303 (1980b):17–19.

Frisch, R. E., A. von Gotz-Welbergen, J. W. McArthur, et al. "Delayed Menarche and Amenorrhea of College Athletes in Relation to Age of Onset of Training." *J. Am. Med. Assoc.* 246 (1981):1559–63.

Frisch, R. E., G. Wyshak, N. L. Albright, et al. "Lower Prevalence of Breast Cancer and Cancers of the Reproductive System Among Former College Athletes Compared to Non-athletes." *Br. J. Cancer* 52 (1985):885–91.

Frisch, R. E., G. Wyshak, T. E. Albright, et al. "Lower Prevalence of Diabetes in Female Former College Athletes Compared with Non-Athletes." *Diabetes* 35 (1986):1101–1105.

Frisch, R. E., G. Wyshak, N. L. Albright, et al. "Lower Prevalence of Nonreproductive System Cancers Among Female Former College Athletes." *Med. Sci. Sports and Exercise* 21 (1989):250–53.

Frisch, R. E., R. C. Snow, E. L. Gerard, L. Johnson, D. Kennedy, R. Barbieri, B. R. Rosen. "Magnetic Resonance Imaging of Body Fat of Athletes Compared to Controls, and the Oxidative Metabolism of Estradiol." *Metabolism* 41 (1992):191–93.

Frisch R. E., R. C. Snow, L. Johnson, et al. "Magnetic Resonance Imaging of Overall and Regional Body Fat , Estrogen Metabolism and Ovulation of Athletes Compared to Controls." *J. Clin. Endo. Metab.* 77 (1993):471–77.

Gerard, E. L., R. C. Snow, D. N. Kennedy, R. E. Frisch, R. Barbieri, B. R. Rosen. "MRI Quantification of Relative Fatness and Regional Fat Distribution in Young Women." *Am. J. Roentgen* 157 (1991):99–104.

Green, B. B., N. S. Weiss, and J. R. Daling. "Risk of Ovulatory Infertility in Relation to Body Weight." *Fertility and Sterility* 50 (1988):721–25.

Halaas, J. L., K. S. Gajiwala, M. Maffei, et al. "Weight-Reducing Effects of the Plasma Protein Encoded by the Obese Gene." *Science* 269 (1995):543–46.

Hartz, A. J., P. N. Barboriak, A. Wong, et al. "The Association of Obesity with Infertility and Related Menstrual Abnormalities in Women." *Int'l J. Obesity* 3 (1979):57–73.

Hershcopf, R. J., and H. L. Bradlow. "Obesity, Diet, Endogenous Estrogens, and the Risk of Hormone-sensitive Cancer." *Am. J. Clin. Nutr.* 45 (1987):283–89.

Jejeebhoy, S. J. "Evidence of Increasing Natural Fertility in Taiwan." *Social Biology* 30 (1983):388–99.

Kaku, M. "Disturbance of Sexual Function and Adipose Tissue of Obese Females." *Sanfujinka NoJissai* (Tokyo) 18 (1969):212–18.

Kennedy, G. C., and J. Mitra. "Body Weight and Food Intake as Initiation Factors for Puberty in the Rat." *J. Physiol.* (London) 166 (1963):408–18.

Keys, A., J. Brozek, A. Henschel, et. al. *The Biology of Human Starvation*, vol.1. Minneapolis: University of Minnesota Press, 1950:753–763; vol.2:839–840; 850 851.

Longcope, C., S. Gorbach, B. Goldin, et al. "The Effect of a Low Fat Diet on Estrogen Metabolism." *J. Endocrin. Metab.* 64 (1987):1246–50.

MacConnie, S., A. Barkan, R. M. Lampman, et al. "Decreased Hypothalamic Gonadotropin-Releasing Hormone Secretion in Male Marathon Runners. *N. Eng. J. Med.*, 315 (1986):411– 47.

Marshall, F. H. A., and W. R. Peel. "'Fatness' as a Cause of Sterility." *J. of Agri. Sci.* 3 (1908):383–89.

Moore, F. K., H. Olesen, J. D. McMurrey, et al. *The Body Cell Mass and Its Supporting Environment*. Philadelphia, WB Saunders Co, 1963.

Nillius, S. J. Weight and the menstrual cycle. In *Understanding Anorexia Nervosa and Bulimia*, Report of the Fourth Ross Conference on Medical Research. Columbus, Ohio, Ross Laboratories, 1983:77–81.

Nimrod, A., and K. J. Ryan. "Aromatization of Androgens by Human Abdominal and Breast Fat Tissue." *J. Clin. Endocrinol. Metab.* 40 (1975):367–72.

Pasquali, R., D. Antenucci, F. Casimirri, et al. "Clinical and Hormonal Characteristics of Obese Amenorrheic Hyperandrogenic Women Before and After Weight Loss." *J. Clin. Endocrin. Metab.* 68 (1989):173–79.

Perel, E., and D. W. Killinger. "The Interconversion and Aromatization of Androgens by Human Adipose Tissue." *J. Steroid Biochem.* 10 (1979): 623–26.

Prior, J. C. "Luteal Phase Defects and Anovulation: Adaptive Alterations Occurring with Conditioning Exercise." *Sem. Reprod. Endocrin.* 3 (1985): 27–33.

Reichlin, S. Neuroendocrinology. In *Textbook of Endocrinology*, 6th ed., R. H. Williams, ed. Philadelphia: WB Saunders Co, 1982:588.

Scammon, R. E. "The Measurement of the Body in Childhood." In *The Measurement of Man*, A. J. Harris, C. M. Jackson, D. G. Paterson, eds. Minneapolis: University of Minnesota Press, 1930:174–215.

Schneider, J., H. L. Bradlow, G. Strain, et al. "Effects of Obesity on Estradiol Metabolism: Decreased Formation of Nonuterotropic Metabolites." *J. Clin. Endocrinol. Metab.* 56 (1983):973–78.

Siiteri, P. K. "Extraglandular Oestrogen Formation and Serum Binding of Estradiol: Relationship to Cancer." *J. Endocrinol.* 89 (1981):119–29.

Siiteri, P. K., and P. C. MacDonald. "Role of Extraglandular Estrogen in Human Endocrinology." In *Handbook of Physiology*, Section 7, v. 2, Part I, S. R. Geiger, E. B. Astwood, and R. O. Greep, eds. New York: American Physiology Society, 1973:615–29.

Snow, R. C., R. L. Barbieri, and R. E. Frisch. "Estrogen 2-hydroxylase Oxidation and Menstrual Function Among Elite Oarswomen." *J. Clin. Endocrinol. Metab.* 69 (1989):369–76.

Strauss, R.H., R.R. Lanese, and W.B. Malarkey. "Weight Loss in Amateur Wrestlers and its Effect on Serum Testosterone Levels." *J. Am. Med. Assoc.* 254 (1985):3337–3338.

Tanner, J. W. *Growth at Adolescence*, 2nd edition, Oxford, Blackwell (1962).

Van der Spuy, Z. M., P. J. Steer, M. McCusken, et al. "Outcome of Pregnancy in Underweight Women after Spontaneous and Induced Ovulation." *Br. Med. J.* 296 (1988):962–65.

Vigersky, R. A., A. E. Andersen, R. H. Thompson, et al. "Hypothalamic Dysfunction in Secondary Amenorrhea Associated with Simple Weight Loss." *N. Engl. J. Med.* 297 (1977):1141–45.

Vissac, B., B. Pereau, P. Mauleon, et al. "Etude du Caractere Culard: IX: Fertilite des Femelles et Aptitude Maternelle." *Ann de Genetique et de Select Animale* 6, no.I (1974):35–48.

Wall, S. R. and D. C. Cummings. "Effects of Physical Activity on Reproductive Function and Development in Males." *Sem. Reprod. Endocrinol.* 3, no.1 (1985):65–80.

Warren, M. P. R. Jewelewicz, I. Dyrenfurth, et al. "The Significance of Weight Loss in the Evaluation of Pituitary Response to LHRH in Women With Secondary Amenorrhea." *J. Clin. Endocrinol. Metab.*, 40 (1975):601–11.

Wheeler, G. D., S. R. Wall, A. N. Belcastro, et al. "Reduced Serum Testosterone and Prolactin Levels in Male Distance Runners." *J. Am. Med. Assoc.* 252 (1984):514–16.

Wyshak, G., and R. E. Frisch. "Evidence for a Secular Trend in Age of Menarche." *N. Engl. J. Med.* 306 (1982):1033–35.

Wyshak G., R. E. Frisch, N. L. Albright, et al. "Lower Prevalence of Benign Diseases of the Breast and Benign Tumors of the Reproductive System Among Former College Athletes Compared to Non-athletes." *Br. J. Cancer* 54 (1986):841–45.

Acknowledgments

The research on the reproductive ability of the runners and swimmers at Harvard University and the Alumnae Health Study was under the auspices of the Advanced Medical Research Foundation, Boston. The research on Magnetic Resonance Imaging was supported by the National Institute of Child Health and Human Development, National Institutes of Health, Grant #HD23536. The author is grateful to the John Simon Guggenheim Memorial Foundation for the Fellowship to study biological determinants of female fecundity, 1975–1976. The research on the quantification of body fat by Magnetic Resonance Imaging was supported by a grant from the National Institute of Child Health and Development, National Institutes of Health.

Glossary

Adipose tissue Body fat consisting of the fat cell, the adipocyte, and connective tissue, the stroma.

Amenorrhea Absence of menstrual cycles, hence infertility.

Androgens Steroid hormones secreted by the testis and the adrenal glands.

Anovulatory Absence of ovulation in a menstrual cycle, hence infertility.

Dyzygotic twins Nonidentical twins resulting from two separate ovulations.

Estrogen Steroid hormone (particularly estradiol) secreted by the ovary, which regulates the development of the uterus and the breasts and other tissues of the female reproductive tract. Estrogens also regulate the characteristic deposition of female fat in the hips, thighs and breasts.

Follicular phase Period of the menstrual cycle up to ovulation during which there is growth of a follicle containing an egg (the ovum) and rapid growth of the lining of the uterus (endometrium). This period normally lasts about 14 days.

FSH (follicular stimulating hormone) Hormone secreted by a part of the brain (hypothalamus) that controls the release of follicle-stimulating hormone (FSH) and luteinizing hormone (LH) by the pituitary gland.

Hypothalamus Part of the brain (the diencephalon) that controls reproduction and other basic functions such as food intake, temperature and control of the emotions. Among other functions, the hypothalamus produces and secretes releasing hormones that control release of pituitary hormones.

LH (luteinizing hormone) Hormone that together with FSH, stimulates estrogen secretion; the surge of LH in the middle of the menstrual cycle stimulates ovulation. LH then controls the transformation of the ruptured follicle to become the corpus luteum (yellow body), which secretes both estrogen and progesterone.

Luteal phase Period of the menstrual cycle after ovulation; the lining of the uterus (endometrium) prepares for implantation of the fertilized egg. If this does not occur, there is shedding of the lining of the uterus and menstruation occurs, normally in about 14 days after ovulation.

Menarche First menstrual cycle. This cycle can often be without ovulation, and the cycles following may be irregular for one or two years.

Monozygotic twins Identical twins resulting from one egg splitting into two parts early in development.

Pituitary gland Secretions of the pituitary gland, controlled by the releasing factors of the hypothalamus, regulate other endocrine organs of the body, including the ovary and the testis.

Puberty General term covering the period of time of the rapid growth of the adolescent growth spurt and development of the secondary sex characteristics before menarche in girls.

Testosterone Androgen produced by the testis. Testosterone regulates the development of male genitalia and male characteristics of the skeleton and muscular system.

9

A Renewable Resource Considered as Capital

By Nathan Keyfitz

Roger Revelle had several careers. The one in which I was associated with him was as Saltonstall Professor and founding Director of the Harvard Center for Population Studies, that occupied him from 1964 to 1978. Coming from the very different science of oceanography, he began at point zero in our field, but quickly established himself as a figure that no one in demography and population studies could disregard.

No one but he can claim any part of the credit for his achievements. His work (along with the late David V. Glass) in historical demography produced an important volume long before I knew him. After I joined the Center, I worked on questions quite different from Roger's, involving much narrower issues, and Roger encouraged me in this. Among many other things which I did not work with him was population and environment, and yet it is my aspiration—surely an impertinence—in this essay to say some of the things that Roger might have said.

With his lightning-quick mind and wide-ranging imagination Roger always regarded me as the plodding type, and might well be laughing at my pretense that I could replicate his scientific style. I had a long talk with him at the 1991 NAS meeting in Washington just a few weeks before he died. He was walking with a cane and was somewhat debilitated physically, a little slowed in speech but not one bit slowed in thought.

It is Roger Revelle the demographer-ecologist, and specifically the unconventional demographer, that I will attempt to celebrate here. If there was anyone who lived by Charles Péguy's,

> Il y a quelque chose de pire que d'avoir une mauvaise pensée. C'est d'avoir une pensée toute faite,[1]

it was he. While others were predicting massive death by starvation resulting from population increase, Roger showed that, without appreciable further technical advance, (Revelle 1975) the planet could provide food for 45 billion people—more than 10 times the population at the time of his writing. His discussion of this fits well into Amartya Sen's subsequent revelation that famines in our times have not been due to crises of overall food supplies, but due to loss of entitlements which result from local unemployment or other impediments which cause obstructions to the access of food. When some said that given the availability of food, the world was actually safe for much more population growth, Roger pointed to the greenhouse effect as a danger, however well the population could be fed. Long before he was joined by others, he warned of the warming, and especially of the shift in rainfall patterns that warming would provoke, and that some large heavily-populated areas would be under seawater, while other areas would become desert. Although the possibility of a greenhouse effect with the increase of carbon dioxide in the atmosphere had been talked about for decades, starting with the suggestion of the Swedish chemist Arrhenius (1859–1927), it was Roger who first brought carbon dioxide into the practical world. He raised the alarm more than a quarter of a century ago, persuading the American Association for the Advancement of Science to establish a committee of scientists to evaluate the danger, and wrote articles on it. By the strength of his personality and his persistence in disregard of the disagreement or indifference of others around him, Roger forced early recognition of what we now know to be indeed a threat, uncertain, of course, but so dangerous that we must not disregard the possibility. He was in the vanguard of those who brought it before the scientific community, and through scientists to the public, some years before it would otherwise have come into popular thinking.

If he were writing today, he would probably put additional stress on social aspects and refer to the virtual certainty that any damage from global warming would fall most heavily on the poor, crowded countries of the tropics, while northern countries like Canada and Sweden would, if anything, gain from the warming.

Reading today's news reports, he might well express surprise to find that some of the countries that are likely to suffer most are demonstrating that they are among the least determined to do something about the problem; whereas those who might gain are actually becoming leaders in the appropriation of funds and in research studies of how to check the warming and how to adopt practical measures for cutting down emissions. He certainly would have commented on the surprising fact that the United States, threatened as it is by the spread of its deserts, is actually hanging back from creating any international agreements to curb warming.

The following discourse is my contribution to this memorial to Roger Revelle. The positions expressed have been inspired by Revelle's approach to population study—views which contrast strongly with the bulk of today's professional work in the field. Look up an issue of *Population Studies*, or *Demography*, or *Genus*, or the titles in *Population Index*, or the articles scattered through other journals, and you will find a concentration on technical matters, some of them purely methodological, mostly establishing a particular fact from data on some locality.[2] For Roger, the big questions were: "Will there be enough food to go around?" and " What does increasing consumption of fossil fuel do to the planet's habitability?"

Exponential Growth Is Only Distantly Relevant

We have heard much about exponential growth, and how it cannot go on forever. If population keeps increasing at even 1 percent per year, it doubles in 70 years, multiplies by 1000 in 700 years, by one million in 1400 years, and so on. Same thing with GNP per head, except faster: if increasing at 3 percent per year, it doubles in 23 years, multiplies by 1000 in 230 years, and by one million in 460 years. The product of these two that determines the pressure on the biosphere is an increase of 4 percent per year, and so a doubling in 17 years. For a while it almost seemed as though articles on population were required to start with such calculations in order to arouse their readers' interest. Though such assertions seem strong, they do not greatly alarm us. That is partly because we are now used to them and partly because nothing increases at a constant speed for even 10 years, let alone several hundred. Even if steady increase were to continue, there would be no sudden change, no dramatic emergency, if exponential increase was the force operating. Each period would always be just the same multiple of the preceding. No reason for panic; if we do nothing to fix things this year, the situation will not deteriorate any more than it did last year. In any case, the fixed rates will not continue; the rate of population growth is already declining, the economy is dematerializing.

This paper sketches a viewpoint that seems to fit better with the approach that, as humans, we adapt to the planet's response to our activities on it. Rather than asking, "What happens if such and such continues at its present rate forever?" instead, we ask two questions:

I. if the resource is nonrenewable, "What happens as population keeps growing when it has appropriated all of a fixed resource?" and

II. if the resource is renewable, "What happens with supplies becoming available each year for just so many people, if annual withdrawals start even slightly to exceed the amount of those annual supplies?"

The first of these is the effect of rigidity, of unwillingness to change, on the part of human agents as they encounter the finiteness or the fixity, of a nonrenewable resource. The second concerns renewable resources, that, despite their name, seem to impose their finiteness more clearly than nonrenewable. We will see how at the end a small change can produce a sudden and calamitous result. This will most often apply when a social entity confronts and lives off a biological structure—say the world's fishing fleets vis à vis the fish in the oceans; humans vis à vis the world's forests; the human population vis à vis the planetary soil and fresh water. Or it may be the human population of the globe living off the aggregate of all of these. The model has unexpectedly wide application; I will even show how it can be applied to tourists in Italy.

But first, we explore the simpler case of a nonrenewable resource, which is to say one fixed in total; paradoxically, the nonrenewable resources on which we depend are less threatened than the renewable ones. The metallic minerals are called nonrenewable, and yet we are never likely to come to the end of them; at the worst, their prices will rise as they become scarcer, but even that has not been happening—the long term trend of nonfuel mineral prices is downward. Thus we can make the treatment of this part brief; in fact, I have only included it in this paper at all to underline the peculiar character of renewable resources to which I will come in the main body of the paper.

I. The Resource Fixed in Total

The Philippine Lowlands

One example of the exhaustion of a fixed resource is the closing of a land frontier. In the Philippines, until contemporary times, there was enough land on the lowland plains for each new generation to establish new villages and make a living the old way. As long as there was free land, the population growth did not at all lower living standards. The population simply kept spreading out. But when the lowlands had become full with the same style of cultivation, practiced people had to settle the hilly forest country higher up. The result is familiar from accounts elsewhere—deforestation, soil erosion, floods descending to the lowlands. After the fact, it came to be realized that the forest, among many other uses, had the function of holding back the rainfall and releasing it gradually to the plain below.

Thus, at one and the same time, came the several phases of crisis, all starting with the exhaustion of good lowland terrain. People could not find farms like those their parents tilled, so many moved to the less fertile hills, with resulting destruction of the protective vegetation. If the peasants had been more adaptable, changing their style of cultivation to make more intensive use of the lowlands, or if they had moved up the hills and left some of the forest, applying the technology of building terraces and other means of holding the soil, they might have been able to continue to live where they were. If they had reduced their birth rate and had been more flexible with regard to their ways of making a living, with some effort and some capital, rural industries could have been established. The surplus population could have, therefore, made a living in the lowlands, thus avoiding the discontinuity of their having to migrate into the hills or into the cities. The discontinuity at the point where the lowlands were full arises from a certain rigidity of the old traditional ways of making a living.

The peasants have now begun the one painless, costless, and sure solution—limiting the size of their families—but they are still increasing much too fast for the land and the jobs available.

Of course, the response to "filling up"—full appropriation of a capital item, land in this case—is culture dependent. The Philippines showed a relatively private exploitation of land and relatively unadaptable technology so that there came a breakpoint, a mutation, in the pattern of human settlement as the population passed a certain point. Yet, in other places (Java is an example discussed below), the filling up was countered by smaller plots and other rearrangements, so that peasants gradually became poorer, and no discontinuity appeared.

By early tradition the Javanese village in principle redivided its rice lands afresh in each generation, with each household being given an area proportional to its needs—needs being defined as proportional to the number of members, including children. Geertz called the process "shared poverty" (Geertz 1963). Incomes tended to fall steadily, with no crisis at any point. That avoids the sudden crisis one finds in other cases cited here, but the incentive to have many children, along with the fixed technology, ensures a steady decline in incomes. In fact, in recent decades the situation has been changed by new technologies in agriculture, the growth of industry on the island, and the spread of birth control.

The hard times of the United States in the 1880s and 1890s coincided with, and could have been caused by, the closing of the American frontier, but Americans had other resources to fall back on, and they were adaptable enough to change. With the end of free land in the West, the stream of immigrants turned north into Canada to exploit the virtually boundless plains of Saskatchewan. That could be done as long as the market for wheat remained strong, which was until the 1930s.

Kenya

The simple filling up of a territory is exemplified also in Kenya. Kenya has always seemed empty; however much the population grew there always seemed to be space for it. Now, it suddenly appears that this is ceasing to be the case. With the kind of agriculture practiced, the population has reached a size that occupies essentially all of the territory. Without the possibility of being able to extend the boundaries of the areas currently under cultivation, Kenya is coming close to a crisis. Only intensified agriculture or industrial development can forestall it.

Burma

At the end of three wars, Britain established a colonial regime in Burma in 1885 and centered its rule in the new city of Rangoon, a port far south of the capitals of ancient Burma, in the thinly inhabited fertile delta of the Irrawaddy. Through the stimulus of private land ownership and tax incentives, colonial Burma became in the course of a few decades a major supplier to international rice markets. By the time World War II broke out it was exporting over three million tons of rice annually and was reputedly one of the richest countries in Asia.

Independence came in 1948, at which time the population was 17 million (according to the United Nations 17,832,000 in 1950 and increasing at 400,000 a year). By 1980 population had increased to 33,800,000, while rice exports had gone down to 674,000 tons (1981 figure according to the *Statistical Abstract* of the United States). The change suggests the classic process of a technologically immobile population climbing up on its food supply. Since 1980 there has been some motion, with improvement in cultivation techniques and the advent of some secondary industry. Whether the improvement is rapid enough and the decline of births sufficient to forestall a crisis when the onetime export surplus is all consumed remains to be seen.

Another example of a population climbing up on an inextensible resource is an underground water reservoir. The Ogallala reservoir in the American West is essentially a fixed (though very large) volume of fossil water, added to, indeed, but only over geological time. When the users come to the end of any such resource—land or water—even if the population is not increasing, a crisis develops unless the society can adapt.

By contemplating these and other examples of an expanding population with fixed culture, living on fixed land, one can list at least five possibilities of escape once land of that quality has filled up:

1. People can settle on poorer land, as Ricardo and the classical economists after him say.

2. People can emigrate or reduce their birth rates, as Kingsley Davis (1963) points out.

3. They can have smaller plots, and be satisfied to be poorer, as the Javanese were observed to do by Clifford Geertz (1963) in the context of shared poverty.

4. They can have intensification, as it is called nowadays, with better techniques used to raise the same or more crops on the old land, as important cases cited by Ester Boserup (1981) show.

5. They can learn to do new things not related to agriculture, though this depends on increased crops produced with the same agricultural labor or less, for which through the market the nonfarm workers can exchange the houses or furniture they build, the clothing they weave, etc. I observed increased agricultural productivity in my second study of Balearjo in East Java (Keyfitz 1985) and a very large increase in secondary industry. In fact this is a variant of the classical model of economic development (Lewis 1954).

All these (and undoubtedly other) possibilities should be seen as responses to the crisis that comes with the exhaustion of good free land. Sometimes the crisis will be more sudden, as on an island with uniform land quality, sometimes more gradual, as where there is a long slow gradient downward of land quality.

So much for the resource fixed in total size, with a culturally rigid population dependent on it. I have had nothing novel to say on it, and only reviewed some cases to illuminate by contrast the main subject of this paper, which is the role of the renewable resources. What the two cases have in common is that the crisis comes on faster than exponentially, but otherwise the mechanism is quite different.

II. The Renewable Resource Capable of Generating a Fixed Income

Income and Capital

By definition, a renewable resource is capable of producing utility year by year without limit of time. This definition applies to extraction of firewood from a forest, to fish from the sea, to crops from the soil. Each of these is like wealth on whose interest we might be living. And because each of them is finite it follows that if our scale of living steadily increases, we reach a critical point sooner or later, and then suddenly find ourselves in serious trouble.

In the light of the following argument the term "renewable" used without qualification is thoroughly misleading. We need a word that means "indefinitely renewable, provided withdrawals remain under a certain upper limit." Such a word is lacking, and yet there is no example of a resource renewable without limit. It is best to think of the nonrenewable resource as having an upper limit to withdrawal in total; a renewable resource as having an upper limit to annual withdrawals. The typical renewable resource is biological, a living entity, while the nonrenewable is inorganic. It is the former that presents the more serious problems.

Arithmetic will exemplify the renewable case. A rentier has a capital of $1,000,000; he has invested it safely so as to bring 5 percent interest indefinitely; inflation does not exist.[3] He has an income of $50,000 per year, that at the start is ample; he draws only $30,000 and adds the remainder to the capital. But his family is increasing, and his expenditures on their behalf increase year by year, so each year he has to draw $1000 more than he drew the year before. The first year he draws $30,000 and adds $20,000 to his capital; the second year he draws $31,000, etc.

In Table 1, we suppose that the interest for the preceding year is credited as of January; that our rentier takes out of the bank on January 2 his expenses for the year ahead, and what is left in the account earns 5 percent interest. The first year there is no interest income on which the rentier can live, and Table 1 supposes that the amounts given in the top row come out of some other source than the original capital or the interest that it earns. During the 10 years that the table portrays the fund does well; the considerable amount of income that is not spent being accumulated. In this, the bank is more generous to him than nature is.

In the nature of interest one can draw only at the end of the period of accumulation, that we are here taking as one year. Hence there is no provision out of this capital for the first year; we are assuming that the income first becomes available on January 1 of year 1, and the year's expenditure can, at that moment, be withdrawn from the bank. No explanation is offered or

TABLE I

Accounts of a Rentier with Fixed Capital and Rising Expenditure: the First Ten Years. (The Footnote Gives the GAUSS Code for This and the Subsequent Two Tables.)

All action taken as occurring on January 1 of stated year; interest on saving starts to accumulate January 1 of the following year.

Year	Capital	Income	Expenditure	Saved
0	1000000	50000	30000	20000
1	1020000	50000	31000	19000
2	1039000	51000	32000	19000
3	1058000	51950	33000	18950
4	1076950	52900	34000	18900
5	1095850	53848	35000	18848
6	1114698	54793	36000	18793
7	1133490	55735	37000	18735
8	1152225	56675	38000	18675
9	1170899	57611	39000	18611
10	1189511	58545	40000	18545
11	1208056	59476	41000	18476

Table 1 and the later tables are generated by the recursion formulas:

$K(T) = K(T-1)+S(T-1)$
$Y(T) = R+K(T-1)$
$E(T) = E(T-1)+ 1,000$
$S(T) = Y(T)-E(T)$
$K(T+1) = K(T)+S(T)$
etc.

where $K(T)$ = Capital in year T,
   $Y(T)$ = income in year T,
   $E(T)$ = expenditure in year T,
   $S(T)$ = saving in year T.

GAUSS program to implement the formulas:

```
>>;format /m1 /rd 10,0;
n=150;i=.05;d=1000;
z=zeros(n+1,1);cap=z;inc=z;expe=z;avail=z;sav=z;
inc[1,1]=50000;expe[1,1]=29000;cap0=1000000;cap[1,1]=cap0;cap[2,1]=cap0;
@cap[t,1] is the capital at the start of the (t)th year@
@inc[t,1] is the income at the start of the (t+1)th year@
@expe[t,1] is the expenditure at the start of the (t+1)th year@
@all expenditures are made at the end of the year and interest received then@
    t=1;do while t<n; t=t+1;
    inc[t,1]=cap[t-1,1]*i;
    expe[t,1]=expe[t-1,1]+d;
    cap[t+1,1]=(cap[t,1]-expe[t]+inc[t,1]);
    endo;
tt=seqa(0,1,n+1);tt=tt-1;sav=inc-expe;
output file=out reset;
"Table 1."
"     Year     Beg cap   Avail inc     Expenditure    Saved      ";
y=tt~cap~inc~expe~sav; y=trim(y,1,0);y[1:12,.];
"Table 2."
"     Year     Beg cap   Avail inc     Expenditure    Saved      ";
v=seqa(1,10,n/10);y[v,.];
"Table 3."
"     Year     Beg cap   Avail inc     Expenditure    Saved      ";
y[90:130,.];
end;
```

T A B L E 2

Rising Expenditure Versus Fixed Capital, Resulting First in Accumulation, Then
Accelerating Consumption of Capital. All Action Occurs on January 1.

Year	Capital	Income	Expenditure	Saved
0	1000000	(50000	30000	20000)
10	1189511	58545	40000	18545
20	1371403	67681	50000	17681
30	1542541	76303	60000	16303
40	1696537	84108	70000	14108
50	1823211	90609	80000	10609
60	1906337	95031	90000	5031
70	1920050	96141	100000	–3859
80	1823127	91971	110000	–18029
90	1549858	79386	120000	–40614
100	995512	53386	130000	–76614
110	–6845	6007	140000	–133993
120	–1723290	–75451	150000	–225451
130	–4577925	–211226	160000	–371226
140	–9246729	–433578	170000	–603578

T A B L E 2 A

Year	Beginning Capital	Available Income	Expenditure	Saved
0	1000000	50000	30000	20000
10	1000000	50000	40000	10000
20	1000000	50000	50000	0
30	950622	48056	60000	–11944
40	753074	39084	70000	–30916
50	319420	18844	80000	–61156
60	–490566	–19355	90000	–109355
70	–1900390	–86180	100000	–186180
80	–4266300	–198632	110000	–308632
90	–8156120	–383809	120000	–503809
100	–14474914	–684902	130000	–814902

needed for the numbers in the first line of the table. All the other lines are
arithmetically consistent. Thus for year 4, 5 percent of 1,058,000 is 52,900;
the saving is this number minus the 34,000 that we assume is to be spent, or
18,900; this last number is added to the capital of 1,076,950 to obtain the
1,076,950 with which the next year starts, etc. To keep the accounts simple,

Rising Expenditure Versus Income Available from Fixed Capita.

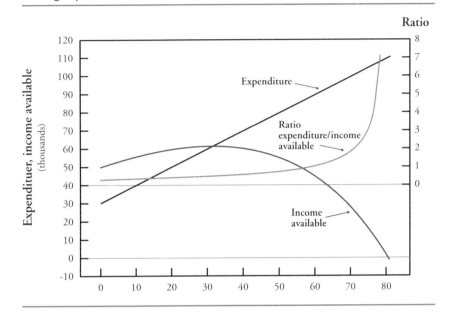

the bank allows no interest on the saving from January to the end of the year in which it takes place.

Table 2 is the same as Table 1, except that it shows only the state of the accounts on January 1 for every 10th year. The capital reaches a maximum of $1,925,302 in the 67th year (not shown), and then starts to decline, very slowly at first, not approaching the million mark again until the 99th year. By this time it is going down very quickly, and is exhausted by some time in the 109th year, only 10 years after passing one million on the down path. (see Figure 1)

Finally, Table 3 is an enlarged view of the crucial years where signs change, shown by Table 2 to be between 100 and 120. That and the chart should eliminate any doubt about the foolishness of drawing down capital.

The table goes on to show the disastrous result of increasing expenditures and financing the increase by borrowing at the same 5 percent interest. (This banker is generous in lending at the same rate as he pays.) It only takes a brief 7 years until the rentier's capital passes the minus 1 million mark, which is to say becomes a debt equal to the amount of the capital initially assumed.

Many other variants of the model are plainly possible. William Reed[4] points out that my model would collapse sooner and more dramatically if I

Rising Expenditure Versus Fixed Capital in the Phase of Accelerating Consumption of Capital.

Year	Capital	Income	Expenditure	Saved
100	995512	53386	130000	−76614
101	918899	49776	131000	−81224
102	837674	45945	132000	−86055
103	751619	41884	133000	−91116
104	660503	37581	134000	−96419
105	564084	33025	135000	−101975
106	462109	28204	136000	−107796
107	354313	23105	137000	−113895
108	240419	17716	138000	−120284
109	120134	12021	139000	−126979
110	−6845	6007	140000	−133993
111	−140838	−342	141000	−141342
112	−282180	−7042	142000	−149042
113	−431222	−14109	143000	−157109
114	−588331	−21561	144000	−165561
115	−753892	−29417	145000	−174417
116	−928309	−37695	146000	−183695
117	−1112003	−46415	147000	−193415
118	−1305419	−55600	148000	−203600
119	−1509019	−65271	149000	−214271
120	−1723290	−75451	150000	−225451

had allowed the family of the rentier to grow exponentially, which is what a fixed rate of increase implies. Using my initial ratios he finds the capital exhausted in as little as 29 years.

Rate of Interest on Overdrafts

Our rentier will be lucky to find a bank that will lend at the same interest rate that it pays out; unless the bank does that, the situation will be worse than that portrayed. Analogous to this, the ecosphere does indeed allow us to borrow from our capital account; we can use up many of the forest or the fishery reserves, and then we can pay back the loan by giving the resource a rest until it restores itself. What rate of interest would nature exact from us when we do that?

I suspect that the interest charge on overdrafts would be great, and estimating that is not easy. However, a formal statement can still be made. We

Capital Accounts on Two Models: the First Compounding, i.e. Allowing Interest on the Interest, the Second Allowing Only Simple Interest on the Original Capital Less What is Spent from it.

	Compound interest			Simple interest	
Year	Capital	Interest	Year	Capital	Interest
0	1000000	50000	0	1000000	50000
10	1189511	58545	10	1000000	50000
20	1371403	67681	20	1000000	50000
30	1542541	76303	30	950622	48056
40	1696537	84108	40	753074	39084
50	1823211	90609	50	319420	18844
60	1906337	95031	60	−490566	−19355
70	1920050	96141	70	−1900390	−86180
80	1823127	91971	80	−4266300	−198632
90	1549858	79386	90	−8156120	−383809
100	995512	53386	100	−14474914	−684902
110	−6845	6007			
120	−1723290	−75451			
130	−4577925	−211226			
140	−9246729	−433578			

should think of the total income—firewood or fish or whatever—that we would have over a period of years between the time when the resource started to decline and when it was able to fully recover. The sooner we recognized that the resource was overused and correspondingly reduced our withdrawals, the less severe would be the dip, and the shorter the time to recovery. However deep or shallow the dip, we would compare the harvest over the period of less-than-normal harvest with what we would have harvested if the resource had remained intact, and we drew only the sustainable yield.

Knowing Where One Stands—Advantage of the Rentier Over the Exploiter of the Resource

Bearing in mind that, even under normal circumstances, harvests vary from year to year, it is impossible to determine the exact moment when a resource itself begins to yield to excessive pressures. Even in retrospect, it is not easy to identify the effects of the strains upon the conditions.

Here is an extremely important difference between the rentier and the renewable resource. The person living off a fixed income receives regular statements from the bank, and his income is expressed precisely in terms of money. He knows just how much he has, and in particular, knows exactly at what point during the increasing withdrawals the capital starts to run down. We all talk of sustainability, but no one can define it in terms that will enable it to be unambiguously identified in the real world.[5] For many of our activities, no one can know at what point and in what respects they become unsustainable, nor what the penalty (interest) is when we make withdrawals beyond that.[6] It is this ignorance of the moment when we cross the fateful boundary into consumption of our renewable resource capital that makes the problem so treacherous, and the establishment of monitoring arrangements so indispensable.

Where Saving Does Not Add to Capital

The above calculation supposed that the resource not consumed was added to the capital. An alternative simplification would be to suppose that the capital remains constant, at least in that the unspent part would not accumulate.

So we modify the model to eliminate accumulation of interest not withdrawn. We will say that the total capital on which interest is drawn has a maximum of $1,000,000 and there is no advantage in consuming less than the full $50,000. The GAUSS program is modified simply by setting the capital back to $1,000,000 whenever it is greater than $1 million. With this modification we calculate afresh Table 2A of this paper.

The table is not very interesting for the first 20 years, its rows being essentially the same. Expenditure becomes equal to income in the 20th year, and from then on the decline comes much sooner than it did for our first model. What I did not appreciate before starting was how great the difference would turn out to be: 110 years with compounding interest, 55 years with simple interest. (For brevity I use the word "compounding" to mean accumulation with interest of unspent interest; "simple" means that unspent interest is lost.) Figure 2 and Table 3A compare the two models, and quantify the handicap of not being able to increase the capital beyond the fixed maximum harvest.

The essential problem of land management is to maximize the interest derivable from a given piece of territory. Suppose that we have at our disposal a small corner of land that though barren could still support trees. Sunlight falling on that bare terrain is either reflected into space and so off the earth, or merely raises the temperature of the atmosphere. Either way, it is lost forever. In contrast, the energy in sunlight falling on trees is captured for future use as protection against erosion, for storing carbon, or for fuelwood. If suitably handled, from maturity onwards it draws simple interest forever.

FIGURE 2

Comparison of Two Models for a Renewable Resource.

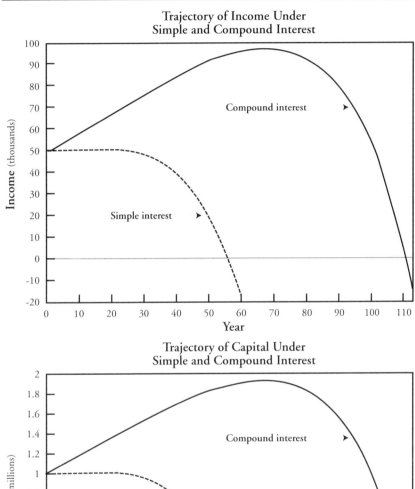

Trajectory of Income Under
Simple and Compound Interest

Trajectory of Capital Under
Simple and Compound Interest

Knowing the Moment When Expenditure Exceeds Current Income

One has to emphasize this uncertainty of sustainability, its intangibility and its dependence, among other things, on future developments in science and technology. Is our present use of tropical hardwoods sustainable? We know that if we continue to cut at current rates, the last natural forests will be leveled within a very few years. Does that bring to an end wooden furniture, wooden forms for cement buildings, etc.? Not if we could have quick growing species that produced a suitable substitute on less land.

Yet to say that is to suppose that those forests do nothing else for us. In fact the forests do many things: they help control the global temperature, provide accommodation to indigenous populations, are the home of hundreds of thousands of species—many not yet even identified—for many of which we could possibly find later use, on many of which we may now be dependent in indirect ways without knowing. Thus, those forests are providing us with portions of our incomes we do not know well enough even to specify, let alone measure.

Insofar as the renewable resource is not only organic but a complex living biological entity, it is more likely to serve as a multi-purpose resource than a nonrenewable resource in our ecological support, since it is more complex than a nonrenewable resource, such as copper or platinum or even oil.

As an example of the latter, consider oil. Suppose that existing supplies are adequate for the next 100 years at present rates of use. At that time, the remaining oil would be so dispersed, and to find other sources would be hopelessly expensive. Our civilization based on oil would then come to a halt. That makes our present use unsustainable, let alone our use in a continuation of the rapidly rising curve that has been traced since the beginning of this century.

But what if, in the next 20 years, someone discovers an inexpensive and safe way of extracting energy by fusion, so that we could move within the following 20 years to a hydrogen-based economy? That would end air pollution and global warming alike, and would anticipate the oil crisis by a clear margin of 60 years. If we have assurance that fusion could avoid this degradation, that therefore, what we are now doing is perfectly sustainable; it can be seen as just one phase in the set of transitions from wood, to coal, to oil and gas, to a hydrogen-based economy.

Nothing similar can be said of tropical forests. The difference between oil and forests is that we come closer to being able to say what purpose the oil serves than what the forest does for us.

So, at one and the same time, I am asserting that while it is vital that we know the exact moment when sustainability is threatened, it is extremely difficult to ascertain that moment, even for a single-purpose resource, let alone for one that serves many purposes. One has to know or guess at what the physical givens are, as well as at what the political changes might be. Further, the most difficult factor to determine of all is the rate of advance in the technology that will be providing the alternative solutions for renewable and nonrenewable resources alike. But, on top of all of these, in the case of renewable resources that are living biological entities, we do not know, and may never know fully, just what those entities are doing for us, which is to say what income they are providing us, let alone when the income is starting to decline.

Other Resources That are Renewable—Within Limits

The rentier model here suggested has remarkably wide application. As another example of a renewable resource, think of Italy's ancient monuments. These ought to be able to bring in a steady flow of tourists, and hence a steady flow of income, indefinitely, forever. Let us see. We now have a marvelous worldwide air network, that has expanded tourism manyfold, and that will undoubtedly continue to expand, and be capable of bringing even more tourists.

In the last few years a new motif has appeared in the brochures mailed out by travel agencies—"unspoiled" has become the important feature that draws tourists. The search for ever new places is on—Mauritius, the Maldives, the Australian coral reef and others that were never thought of by the masses of tourists a decade or two ago.

This is partly because the old places have become banal, but equally because they have become contaminated in one way or another. The countless thousands of tourists have, through their visits, brought desolation to the north shore of the Mediterranean and other more accessible tourist spots. Are a country's tourist sites a renewable resource, one that would be sold to tourists year after year forever without losing value? Yes, but not after tourism becomes as big as it is now becoming. Unless there is some change, each spot is viable for only a few years, in the course of which it is trampled over, eaten away by fumes, taken away piece by piece by souvenir hunters, and so loses its value for attracting other tourists.

On a finite globe at the rate of expansion of the 1980s all tourist spots could well be destroyed as such within a few decades, and the tourist phase of history will close. Here is one way in which a technical, economic development, in this case the world air network, as used for recreation, undercuts itself. The rentier model is suited to this case just as well as to income generated by forests or fish. On the other hand, the capacity of the atmosphere to absorb emissions without excessive warming had best be seen as a nonrenewable resource, at least over a brief interval of geological time like the next century or two.

The so-called renewable resources quickly become nonrenewable once users start to harvest more than the annual long term sustainable crop, i.e., the crop that can be produced without damage to the resource base. This applies to fisheries, soils, forests, even tourist spots. With an increasing population and an expanding economy living on a fixed income of fish etc., the moment the whole of the annual output is consumed starts an accelerating decline of income, and of the underlying resource. In this respect the analogy to a rentier living on coupons is exact.

Where it fails to be exact is that the rentier knows exactly where he stands at any moment; his banker can tell him to the last dollar. The resource users are in a much more difficult position, spending without knowing just what the sustainable level is. That is partly because the resource is hard to measure at any one time, though the precision of statistical data available can usually be increased by more effort put into measurement. What is not so easily remediable is unanticipated year-to-year, decade-to-decade, and century-to-century variation in the annual return.

Given the acceleration once the decline starts (whose results we see now in the commercial fish of the world's oceans) and the uncertainty as to where we stand at any moment, the only prescription possible is extreme caution in allowing consuming populations and economies to grow towards the point where use equals output.

References

Boserup, E. *Population and Technological Change.* Chicago: University of Chicago Press, 1981.

Davis, K. "The Theory of Change and Response in Modern Demographic History." *Population Index* 29, no. 4 (1963):345–66.

Geertz, C. *Agricultural Involution: The Process of Ecological Change in Indonesia.* Berkeley: University of California Press, 1963.

Keyfitz, N. "An East Javanese Village in 1953 and 1985: Observations on Development." *Population and Development Review* 11, no. 4 (1985): 695–719.

Lewis, W. A. "Economic Development with Unlimited Supplies of Labour." *Manchester School of Economic and Social Studies* 22 (1954):139–91.

Revelle, R. "Will the Earth's Land and Water Resources be Sufficient for Future Populations?" In *The Population Debate: Dimensions and Perspectives*, vol. 2. Papers of the World Population Conference, 1974, ST/SEA/SERA/57. New York: United Nations, Department of Economic and Social Affairs, 1975.

Sen, A. *Poverty and Famines: An Essay on Entitlement and Deprivation*. Oxford: Clarendon Press, 1981.

Simon, J. L. *Population and Development in Poor Countries*. Princeton: Princeton University Press, 1992.

U.S. Bureau of the Census, *Statistical Abstract of the United States*. 1994 (114th Edition), Washington, DC

Acknowledgment

My thanks to Robert Dorfman, who meticulously went through the details of my calculations and suggested several improvements.

Notes

1. There is something worse than having a wrong thought. It is having a ready-made thought. (Quoted by Edwy Plenel in *Le Monde* of January 17, 1992, 26.)

2. A few titles taken haphazardly will make my point: Cronk, L. "Wealth, Status and Reproductive Success Among the Mukogodo of Kenya." *American Anthropologist* 93(2):345–360, June, 1991. Washington. DC; Malhotra, A. "Gender and Changing Generational Relations: Spouse Choice in Indonesia." *Demography* 28(4):549–570, November, 1991; Légaré, J. "Infant Mortality Among the Inuits (Eskimo) after World War II." *Genus* XLV(3–4), July–December, 1989; Blum, A., and Monnier, A. "Mortality by Cause in the Soviet Union." *Population* 44(6), November–December, 1989.

3. The most famous expression of the model that follows is Charles Dickens': "Annual income twenty pounds, annual expenditure nineteen nineteen and six, result happiness. Annual income twenty pounds, annual expenditure twenty pounds ought and six, result misery." *(David Copperfield,* Chapter 12)

4. Letter dated 23 July, 1993; quoted with permission.

5. I have not seen anything that goes beyond the much repeated statement of the Brundtland Commission that sustainable development is, "Development that meets the needs of the present without compromising the ability of future generations to meet their own needs…Thus sustainable

development becomes a goal not only for the 'developing' nations, but for the industrial ones as well." Proceeding from this to numbers even in respect of one particular resource involves many problems, and a fortiori for the aggregate of all resources.

6. Erik Arrhenius in conversation makes the general point that stability in a system, in itself desirable, works against our having a warning of the approach of a critical point, and makes the consequence of passing the critical point more violent and more sudden.

10

Malthusian Scenarios and Demographic Catastrophism

By Geoffrey McNicoll

For a distinguished university-wide lecture series at Berkeley in 1969, the invited speaker was the prominent oceanographer, Roger Revelle. No respecter of disciplinary parochialisms, Revelle had already abandoned the marine depths at the Scripps Institution for the muddy sloughs of population policy at Harvard, and he chose "population and development" as the topic of his lectures. Berkeley demographers (I was at the time a graduate student among them) made their displeasure felt at such interloping. The gulf between the natural and social sciences was wide, and population belonged on the social side. Yet Revelle's concern, then and later as Director of Harvard's Population Center, was with bridging that gulf. The need to do so is now conceded by most demographers. If the bridge remains fragile and little used, it is because Roger Revelle had too few emulators.

Introduction

Ecologists and environmentalists record grim stories of land degradation, pollution and species extinctions. Expanding human presence and economic activity presage, at best, the "end of nature," at worst, a thoroughgoing despoliation of the planet. Where human numbers are innocent as cause, they may still enter as effect, magnifying the consequences of natural disasters. In considering policy, the ecologists' stance is curatorial or, in the deeper reaches, "biocentric"—conscious of obligations extending far into the future and ranging across species and ecosystems.

Social scientists, by and large, give only peripheral recognition to such views; some are openly dismissive. They prefer to stress the malleability of human arrangements. They put store in the innovativeness that has freed economies and societies from numerous past predicted limits, finding common ground, in this respect, with the technological optimists. For policy, they are willing to recommend more diligent efforts at "getting prices right" or improving institutional design and performance, but reject more radical remedies for environmental problems as politically infeasible, ethically impermissible (ethics being enlisted on both sides), or simply failing to meet elementary cost-benefit criteria.

Population growth provides a notable instance of this failure of engagement. The standard scenario for the demographic future from a social scientific standpoint is as follows: the already perceptible slowing of world population growth will continue, as will the substantially faster fall in family size. Some time in the twenty-first century, uniformly low mortality and replacement level fertility will be attained and, not long after, growth will come to a halt; thereafter, demographically speaking, nothing much will happen. This is the future spelled out in the medium-variants of the long-run projections made by the UN Population Division and the World Bank, which have world population size plateauing at between 11 and 13 billion in the twenty-second century (United Nations 1992; Bos et al. 1992).

Demographers do, of course, construct alternative scenarios. Indeed, they come cheap. The UN's are the best known: its low and high world population trajectories are often taken as delimiting the "plausible range" of demographic futures. In the most recent version, these are obtained simply by varying the level at which fertility eventually stabilizes from 1.7 children per woman to 2.5. (The same downward mortality trends are assumed in each case.) The resulting population totals for 2100 are 6 and 19 billion; for 2150, 4 and 28 billion (United Nations 1992). These diverse outcomes, however, maintain the presumption of smooth convergence of demographic patterns everywhere to ones approximating those of today's rich countries.

By incorporating this expectation of soon-completed demographic transition, such figures depart from simple extrapolations of current trends, but they remain purely demographic exercises. For example, the transition they depict is presumably brought about in large measure by across-the-board economic development, but the ecological consequences of that development, which themselves may have significant demographic effects, are unexamined. Implicitly, resource and environmental limits remain distant and technological advances are either autonomous or endogenous—available off the shelf at a steady rate or "induced" by the demand for them. Ecologists express doubts on both scores. In the "limits to growth" literature, various candidates for limiting factors in population growth are proposed that would cap population size or its rate of increase, ruling out the more extravagant projections of human numbers. Even fairly modest projected demographic outcomes may be judged to be inconsistent with the welfare levels most people aspire to.

Ecologists are joined by a wider array of scientists in also calling attention to possibilities of disruption of any smooth approach to a zero population growth state. There is a repertoire of environmental "surprises" that have that capability, drawn from earlier global history; there are others, no doubt, as yet undreamed of. Moreover, systems modeling, once seeming to confirm the stability of many processes in the natural world, now reveals instead their unpredictability.

The disjunction between the relative equanimity of social scientific future-gazing and the very different expectations of ecologists and environmentalists warrants more exploration than it has had. Its demographic dimension is an appropriate topic for the present symposium, given Roger Revelle's long-standing interests in both population and natural resources. Revelle stuck to matter-of-fact accounts of population-resource relationships, generally finding reason for a measured optimism about the future. In this essay I shall comment on such accounts and examine why they do not seem to settle the arguments.

Population Ceilings

One of the more familiar terms in popular ecological debate is carrying capacity. For any given base of resources and technology and specified average consumption level there is a ceiling on population size set by one or other essential factor. (Resources include not only material inputs to consumption but also environmental sinks that can accommodate by-products and waste.) Where known resources are expanding or technological capacities advancing, population ceilings may also be rising.

The application to human populations is much the same as for animal husbandry (see Demeny 1988, for a historical account). Depending on assumptions about future input, supplies and technological possibilities, a time-trajectory or upper bound for human carrying capacity can be calculated. Revelle's 1974 and 1976 *Scientific American* articles are straightforward instances of such calculations, reporting potential areas of arable and irrigable land by continent and assessments of average yields that should be attainable. For example, at an average food intake of 4–5,000 kilocalories grain-equivalent per capita per day, Revelle (1974, 168) estimated total carrying capacity for the world to be 38–48 billion persons.

Virtually every assumption in any such calculation can be disputed. To take a well-known instance, the FAO-IIASA study of carrying capacities in the Third World, comparatively sophisticated in its treatment of soil and climate, is seen as highly extravagant in its judgment of potentially cultivable land—envisaging a trebling of the present cultivated area—and extraordinarily modest about (and reticent in stating) dietary allowances of around 3,000 kcal (Higgins et al. 1982). Under a low-technology assumption, the Third World (excluding China), which had a 1990 population of 2.9 billion, could supposedly support 5.6 billion; with modern agricultural inputs this number rises to over 30 billion—three-quarters of it in Africa and South America. (The West African nation of Guinea, with a 1990 population of 5.7 million, supposedly has the land resources, under high input levels, to support 170 million.)

Gilland (1979, 1983, 1984), in stark contrast, argues that the population size ceiling is already very close. He is scathing on the area assumptions of the FAO-IIASA study and would even cut back Revelle's assumed doubling of cultivated area. Allowing a more varied and higher-protein diet (9,000 kcal grain-equivalent), he derives a world-wide carrying capacity of 7.5 billion (Gilland 1983, 206). At the current world average intake of 6,000 kcal, the corresponding population level would be 11.5 billion, which coincidentally happens to be the projected stationary world population under the UN's medium-variant assumptions. At the South Asian average (3,000 kcal), the population can reach 22.5 billion and at the North American average (15,000 kcal), 4.5 billion.

Land areas and soil potential should be determinable as matters of fact. However, differing assumptions about areas assigned as permanent forest cover or otherwise reserved from agriculture (and about land lost to cultivation by settlement or disuse) introduce large variation in results. The extreme case is represented by the FAO-IIASA exercise, where feasibility alone limits the expansion of agriculture: nearly all areas under tropical forests, for example, are converted to cultivated or range land. In a study much cited by environmentalists, Vitousek et al. (1986) offer an alternative way of probing these physical limits. They calculate the fraction of total organic material produced through photosynthesis that is now being appropriated by human activity. The authors estimate that roughly 224 billion tons of organic matter are produced annually as a result of the (mostly solar) energy that is fixed biologically— 132 billion tons on land, 92 billion tons in water. The amount that is currently used directly by humans or used in human-dominated ecosystems is put at 40 billion tons on land (30 percent of the terrestrial total) and 2 billion tons in water (2 percent of the aquatic total). Human "degradation" of the terrestrial environment is judged to have prevented the production of another 17 billion tons, which if added to numerator and denominator raises the 30 percent fraction to nearly 40 percent. Given the projected further doubling of population, the calculation would imply that the "end of nature," at least on land, is indeed close.

The attainable average crop yield assumed by both Revelle and Gilland is about 5 tons per hectare. The current level is 2 tons. In the medium-term future, yield advances are likely to come chiefly from conventional varietal improvement and greater use of modern inputs—in particular, synthetic nitrogenous fertilizer. Smil (1991) estimates that the food contribution of this fossil fuel-derived source of nitrogen already sustains one-third of the human population.

The longer-run technological prospects for exceeding such yields are hard to assess. Expert opinion seems fairly evenly arrayed across a wide spectrum. Technological optimists—for example Avery (1985) and Crosson (1986)— see few difficulties with high rates of agricultural productivity growth continuing virtually indefinitely, given sensible institutional arrangements, and none in supporting UN medium-variant population numbers. Molecular biology and genetic engineering are still in their infancy; the new technologies they will spawn may yet cut the link between food production and arable land, making the rural landscape again look as empty as it must have seemed to the first agriculturalists. While such developments may be far off, the past record of technological change gives credence to the case. Its supporters are not slow to invoke what might be called the horse dung argument: that the supposed new limits to growth will prove no more binding than the limit once predicted for London's growth set by the stench and disposal problems of horse manure.

Technological optimism has to confront ecological pessimism. Many environmental trends are in the direction of degradation or depletion of agricultural inputs, notably soil and water. Resource and environmental constraints are often of a kind that permit over-exploitation, ultimately leading to fall-back in sustainable yield. Levels that in theory could be reached under best-practice technology and policy bear no relation to those that actual practices will deliver.

The sheer range of estimates is often seen as a reason for not taking carrying capacity calculations very seriously. Global population may perhaps be bounded at 7.5 billion or at 40 billion. The implied uncertainty is deemed sufficient to dismiss the exercise as vacuous. Of course, much of the range is attributable to assumptions that are strictly extraneous to the issue. To the extent the differences result from varying assumed levels of consumption per head, we have merely changed the subject to that of human welfare (including the "Benthamite question" of whether any merit attaches to a welfare criterion that weights per capita well-being by population size). To the extent they derive from varying notions of human stewardship of the planet, whether that is seen as an ethical imperative or simply as a value that predictably enters the utility functions of many people once their more elementary needs are securely met, we have changed the subject to that of environmental ethics.

A more fundamental reason for the scant regard accorded limits-to-growth arguments among social scientists is the inaptness of the ceiling metaphor itself. Calculations of global carrying capacity like Revelle's or Gilland's refer to the hypothetical situation in which food consumption is roughly evened out across regions and income groups. Since redistribution in favor of the least well-off regions or strata is in fact fairly minimal, the "ceilings" for any specified consumption standard will be encountered much earlier. The concept of a global ceiling becomes redundant.

What is left are the standard-order problems of low productivity and poverty. The feedbacks that signal most of the limits in question are continually operative in one guise or another. Regional and social heterogeneity govern where they bite; a society's political and cultural makeup, and the international institutional regime in which it finds itself, govern the nature of the responses elicited. Hence, it is not that we are now for the first time approaching an age of limits: aside from some short-lived excursions, we have never left such an age. The analysis of those feedbacks and the design of systems that select and promote appropriate responses are the main stuff of development thinking and policy.

The argument on this point is exemplified in the familiar contrasting readings of Malthus. In the more common reading, based on the first edition of the *Essay*, Malthus is taken to be the prophet of resource crisis, a distant progenitor of the Club of Rome. The claim is not without some basis, given the harsh adjustments to subsistence limits he describes. Yet what fills the increasingly thick later editions of the *Essay* are not accounts of population crashes of the sort seen in, say, the Club of Rome's World3 model (Meadows et al. 1973) but detailed case studies of demographic systems variously responding to resource scarcities. Malthus's interests were in what it is about the constitution of societies that influences their reaction under that kind of stress. This systemic concern, divorced from Malthus's particular moral standpoint, remains an important theme of present-day analysis of population problems.

The adjustment processes that curtail population growth are voluntary ("prudential") or state-encouraged reductions in fertility, emigration (where feasible), and greater attrition through mortality. Responses can also take the form of induced productivity increases, allowing, and often resulting in, deferral of the demographic adjustment.

Hunter-gatherer societies, the prevailing social system over most of human existence, appear to have controlled fertility through long birth intervals (four years among the !Kung), with the effect of holding substantial potential fertility in reserve and able to be drawn on to counter mortality shocks. Subsequent agricultural societies may well have needed those fertility reserves in the face of the more severe disease environments that came with permanent settlement. High birth rates became institutionalized. In relatively few cases—Japan and western Europe are the best documented—societies developed demographically-effective social controls on marriage or household formation that were able to respond to population size considerations. Elsewhere, for the most part, equivalent institutional arrangements, if they ever existed, had long withered. This century's reductions in overall (and particularly peak) mortality were translated directly into dramatic population growth over much of the world.

Reestablishment of low fertility, where this has occurred, has come about less through new social controls devised or emerging in modern societies than through the individualizing process of modern urban-industrial development—which has sheeted home the costs of children to their parents and transformed the roles of the family and the expectations of its members. The case of China in the 1970s, where state pressures played a large part in the halving of fertility, may be the only noteworthy exception. The fertility decline is now nearing "completion" (i.e., fertility is getting close to replacement level) in about half the world's population.

For many observers it has seemed self-evident that all societies would eventually experience the same transition to a modern demographic regime. Confidence in this progression, however, may now be eroding. "Stalled transitions" are increasingly familiar, in which population growth persists at, say, 2 percent annually for decades, absorbing resource slacks (forests, soils, ground water) and productivity increases (based on high yielding crop varieties and synthetic fertilizers and pesticides) without the broad-based social and economic development that is believed to foster fertility decline—or with promising early steps in that direction faltering in civil disorder. From such a situation, an edging up of mortality may be as likely a demographic outcome as lowered fertility.

If we trust the statistics, no country as yet has experienced such a demographic reversion. But the proviso is a substantial one. Adult mortality in particular is hard to measure, the more so under the conditions—including breakdown of government administration and large-scale migrations—that would typically exist if it were greatly increasing. Estimates come with multi-year delays and ample use of imputation based on conventional patterns to fill out missing data. High death rates, moreover, attract proximate explanations (droughts, famines, epidemics) that presuppose a "normal" underlying force of mortality that will be restored in due course.

Region-wide increases in mortality in a roughly homogeneous agrarian society, even if poorly documented statistically, are likely to be evident enough through casual observation. This may not be the case, however, when the process is strongly class specific. Under population growth in a static economy, many families and individuals are necessarily pushed down the economic scale, with those at the bottom becoming increasingly marginalized in terms of access to economic product and health services and exposed to a harsher mortality regime. The process of downward economic mobility and recruitment into higher mortality-risk groups can coexist with apparent relative stability in the overall asset distribution.

These are cases that would be familiar to Malthus. Not so would be the postulated mortality effects sometimes ascribed to "pollution," for example, those in the World3 model. While negative feedback loops are easy to incorporate in systems models, with potentially spectacular effect, it is harder to envisage the empirical situations they seek to depict. Massive industrial pollution in eastern Europe and the Soviet Union, to take a case in point, unquestionably had and will continue to have serious health consequences for much of the population (Feshbach and Friendly 1992). It is probably a factor in the lower life expectancy in the region than in Western Europe. However, its significance for population growth is trivial in comparison to birth control practice. Pollution is a major issue for health, quality of life, and environmental aesthetics and ethics, a minor one for demography. Feedbacks operating through climate and other global systems are taken up below.

Population Collapses

The kinds of adaptations I have been discussing can together be classed as Malthusian processes, although they are not all covered explicitly in the canonical volume—and some would have elicited Malthus's severe disapprobation. Their common feature is that of demographic and economic change at the margin to maintain or reestablish an equilibrium between population and the resource base—through deliberate measures adopted by individuals or (sometimes fortuitously) enacted by societies, or by "natural" feedbacks contained in the biological, social or socio-environmental systems. For present purposes, therefore, the population pressure-induced process of productivity improvement described by Boserup (1965) and others is deemed Malthusian-in-spirit. Malthus expected that such improvement would usually be dissipated in further population increase—an expectation often borne out. Where the gains can be sustained, however, intensive economic growth may be engendered (on the conditions for this, see Jones 1988).

Demographic change need not be marginal. Locally, both migration and mortality can attain massive levels over short periods; regionally, the same theoretical possibility exists, though with reduced feasibility in the case of migration; and globally, catastrophic mortality is at least conceivable and at times has been much more than that.

The "great mortalities" of human history are chiefly attributable to disease. Disease is lifted out of routine experience when new or newly virulent pathogens appear in an unprotected population, such as bubonic plague in medieval Europe or the Euro-Asian diseases in the postColumbian New World. Density of settlement and intensity of contact across human societies have by now virtually eliminated population isolates. Those same features, however, mean that humanity as a whole is potentially a kind of isolate with respect to some pathogens crossing interspecific barriers from animal populations—HIV apparently being one according to the principal hypothesis of its origin—or with respect to mutations affecting the lethality or transmissibility of existing disease strains. There may be a large measure of inevitability about both these processes, just as it was inevitable that the indigenous peoples of the Americas and the Pacific would eventually be exposed to Old World diseases.

The degree to which technological or institutional responses may be able to lessen the mortality impact of such threats in the future is hard to assess. Joshua Lederberg (1988) is skeptical that future victories are assured, arguing that our biomedical successes have instilled a false confidence. "Most people today are grossly over-optimistic with respect to the means we have available to forfend global epidemics comparable with the Black Death of the 14th century." (Lederberg 1988). He draws an arresting parallel between the human condition and a culture of bacteria beset with its own viral parasites, the bacteriophages:

It is not unusual to observe a thriving bacterial population of a billion cells undergo a dramatic wipe-out, a massive lysis, a sudden clearing of the broth following a spontaneous mutation that extends the host range of a single virus particle...Perhaps there are a few bacterial survivors: mutant bacteria that now resist the mutant virus. If so, these can repopulate the test tube until perhaps a second round, a mutant-mutant virus, appears. Such processes are not unique to the test tube. The time scale, the numerical odds, will be different. The fundamental biologic principles are the same (Lederberg 1988).

Famine, despite its prominence in the more alarmist ecological literature, has probably not been a major controlling factor in past population growth. Watkins and Menken (1985) concluded from a detailed simulation study of the issue that: "The only way famines and other mortality crises could have been a major deterrent to long-run population growth when the underlying normal mortality and fertility rates would have led to even moderate growth is if they occurred with a frequency and severity far beyond that recorded for famines in history." Looking particularly at the Indian case, Cassen and Dyson (1976) show that even assumptions of "medieval mortality"—death rates fluctuating around 15–20 per 1,000 in "normal" years, interrupted by famine or disease peaks of twice that level every 15 years—would have only a minor effect on India's projected population trajectory.

The largest famine mortality in recent times, and perhaps in human experience, occurred in China during 1959–61, associated with the disastrous "Great Leap Forward" policy. Estimates of excess deaths lie in the range of 23–30 million (Ashton et al. 1984; Peng 1987)—amounting to 3–4 percent of the population. In the worst-affected province, Sichuan, there were 7.5 million excess deaths, more than 10 percent of the population. For the country as a whole, the depression of fertility, another major consequence of famine conditions, was estimated to be the equivalent of nearly one year's births (Peng 1987). While clearly visible in the national and even Asia-wide time series, the episode was soon to be swamped in effect on the demographic aggregates by the onset of China's secular fertility decline.

Conventional wisdom on the causes of famine has undergone substantial reassessment in recent years, with emphasis shifting from production failure—something that might readily be supposed to have unavoidable demographic consequences—to flaws in institutional arrangements and information flows (and, often magnifying or underlying them, government neglect, incompetence, predation or malevolence). The Chinese famine owed much to the perverse incentives for cadres to inflate crop estimates, blinding the central government to the true supply situation. The 1929–33 famine in the Soviet Union, among the century's greatest (with around 7 million excess deaths in the Ukraine, Northern Caucasus and Kazakhstan), was the product of a ruthless and ill-judged collectivization policy, tinged with what today might be called "ethnic cleansing" (Conquest, 1986). The 1943–44 Bengal and 1974 Bangladesh famines, as interpreted by Sen (1981), were less supply failures than collapses in "food entitlements." Sahelian famines have been linked to harsh procurement schemes, sedentarization policies, and long-running civil wars. Production failures, of course, cannot be wholly absolved through such arguments. It is merely that good policies can find ways of coping with a much larger class of production failures short of heightened mortality.

War in its "conventional" form, like famine, also has a relatively small effect on aggregate demographic change, drastic though its consequences may be in the regions immediately involved. Richardson's (1960) classic study of "deadly quarrels" between 1820 and 1949, ranging from murders to world wars, found the resulting deaths to amount to about 1.6 percent of all deaths in that period. Wright (1942) estimated Europe's war casualties in the early decades of this century at 3 percent of all deaths in Europe over the period. The same percentage applied to worldwide deaths over 1900–90, some 4 billion, would give 120 million, which would probably be an overestimate, but not by a great margin, of the century's war-related fatalities.

Where larger population effects are recorded, explanations tend to also involve the destruction of the social infrastructure of civil administration and rule of law, and reversion to a simpler economic system. In one of the more striking instances, evidence points to a 80–90 percent fall in population associated with the Mongol invasion of Persia in the 13th century and subsequent drawn-out political turmoil—particularly affecting irrigation management (Jones 1988, ch.6; Bowden et al. 1981).

Nuclear war of course has been in a class of its own, with worst-case assumptions of all-out war giving over a billion deaths (World Health Organization 1983) and possible "nuclear winter" (or, in revisionist accounts, "nuclear autumn") scenarios raising that number much further. That threat, if not the risk of nuclear weapon use in more confined conflicts, has abated considerably, but over the century or more covered by the demographic projections we began with there is no reason to suppose it could not reemerge. Biological weaponry may one day present a threat of the same order of magnitude.

What about secular climate change, for several years preeminent among the crowded assembly of popular ecological crises? Here too, the balance of opinion is for a less than catastrophic outcome, although there is enough uncertainty to leave open a significant downside risk. The most publicized climate models suggest a warming over the next century enough to shift grain belts several degrees toward the poles—requiring substantial adjustments in cropping patterns, crop varieties and pest control systems—and to produce many, mostly adverse, changes in natural ecosystems. Schelling (1984, 1992) gives a business-as-usual interpretation of social consequences—though with medium-term cause for worry in poor countries (and note taken of the possibility that an alteration in the Gulf Stream might "freeze Western Europe"). Cline (1992) foresees larger adaptation problems. The models may of course be wrong: as more becomes known about the behavior of nonlinear systems—including the severe theoretical limits they impose on prediction—future possibilities seem to be widening rather than narrowing. Revelle, writing in 1977, remarked that it may never be possible to predict future climates. "It is not unlikely that without any change in such external parameters as the quantity of incoming sunshine or the carbon dioxide content of the air, quite different climatic regimes may succeed each other. The climate may proceed from one state to another because of its internal dynamics in an essentially stochastic manner" (1977, 196). The cautious comment by Flöhn (1986, 147) reflects a now common sentiment:

> Taking into account the similarities between the results of nonlinear dynamics and the available evidence of regional or near-global climatic discontinuities, the possibility of an abrupt evolution of the climatic system, or a shift into another mode, in the foreseeable future should no longer be excluded from scientific discussion. The classical paradigm 'natura non facit saltum' appears to be limited.

For completeness, any catalogue of putative catastrophes that might govern the demographic future should include natural disasters that are unrelated to the demographic past or societal present. Radical climate change, as the above quotations suggest, may belong here. A major collision with a "near-Earth object" (NEO) would be a more exotic case, working through the effect of resulting atmospheric dust on temperature and hence on food production. The aftermaths of major volcanic explosions give some indication of the latter process. Krakatoa in 1883 is thought to have released 6–18 cubic km of solid matter into the atmosphere, largely as dust. Tambora, also in Indonesia, in its 1815 explosion may have ejected several times this amount—causing northern hemisphere crop failures in 1816, the "year without a summer" (Lamb 1988). NEO collisions present a potentially larger source of atmospheric dust and debris. The average frequency of Earth impacts by objects 50 meters or more in diameter (that is, comparable to or greater than the presumed cause of the "Tunguska event" in Siberia in 1908) is estimated to be around one per century. The frequency for objects 1 km or larger—with the potential to produce a "cosmic winter" and mass extinctions—is believed to be one per 100,000 years. Steel (1991) maintains that the most significant risk for humans is "low-level airbursts by 50–500 meter objects." Students of NEOs disagree on whether such events are effectively random or occur in cyclical episodes—a distinction between "stochastic catastrophism" and "coherent catastrophism" (Steel 1991, 266). For some size-range of NEOs, human technology will no doubt eventually devise means of early warning or even ways of diverting the object to avoid collision.

What then are the prospects for human population collapse of the sort that animal ecologists are familiar with—for a "clearing of the broth"? Certainly, the sudden growth in human numbers this century might be said to resemble the "explosion" that would precede such an event in populations of lemmings or flour beetles. The chief objection to the analogy is that technological and institutional innovation in the human case provides routine sustenance for the vastly increased population as well as added resiliency against shocks, and the capacity to anticipate and avert or adapt to many potential dangers. The demographic growth and regime-transition underway are then more analogous to the prehistorical ones which must have accompanied the replacement of hunting and gathering by agriculture as the dominant economic system of human societies. In part, this objection is indisputable: world population is five times its level at the start of the industrial revolution and mortality has never been lower. There remains uncertainty, however, about the scope of the resiliency and adaptive capacity that have been acquired. The 50-year flood is taken care of, so to speak, but we may be defenseless against the 500-year one. Adaptation may limit adaptability. Holling (1986) has argued strongly in this vein with reference to the resilience and stability of ecosystems: "Slow changes...might be so successfully absorbed and ignored that a sharp, discontinuous change becomes inevitable."

In more stylized fashion, this is also roughly the "contradiction" that Fred Hoyle, in a brief excursion from cosmology into demographic speculation, detected in the Malthusian equilibration argument. Far from fertility and mortality eventually coming into equality to complete a demographic transition, Hoyle ([1963], 1986) saw nothing to prevent a continual racheting up of population numbers and further distancing from subsistence limits, as successive obstacles to expansion were overcome by technological and organizational prowess. Increasing systemic complexity, however, would end not in a jostling, high-consumption nirvana like that implied in Simon's *The Ultimate Resource* (1981), but in eventual organizational overload and collapse—destroying the productive base and with it most of the population it supported. In Hoyle's expansive time frame, the game can then begin again: it is, as he put it, a future "in some ways more horrible, in some ways more hopeful, and certainly in all ways less dull" than that of Malthus (Hoyle [1963] 1986, 552).

Realities, Values, and Habits of Mind

It is nearly a century and a half since J. S. Mill replaced the bleak penury foreseen by Smith and Malthus by a vision of a civilized arcadia. Yet we are not much closer to settling the question of how the world's population growth will eventually cease. The partial scenarios and scraps of theorizing and speculation that I have reviewed above give pause to the simple expectation of a succession of national demographic transitions to low mortality and low fertility, especially ones that would be accompanied by high consumption and increasing respect for environmental values. A more variegated set of alternatives is generated, including regionally-specific conditions under which populations grow to levels that at present seem scarcely feasible, and radical fall-backs occur in numbers as a result of natural or manmade disasters.

With such possibilities in mind, I return to the question I started with: what underlies the gulf between social scientific and ecological views of the demographic future, and why the considerable mutual disdain between many of those standing on opposite sides? The explanations appear to involve differences in knowledge, values and mindsets.

Differences in what we know or think we know must have a lot to do with the contrasting viewpoints. In assessing the same hypothetical population trajectory and its ecological consequences, different empirical illustrations come to mind as pertinent analogies and different theoretical structures are brought to bear. A biological training would tend to emphasize behavioral commonalities and continuities across the human species; a social scientific one would stress cultural adaptability and idiosyncrasy. To the extent that the latter are downplayed, ecological degradation will tend to be seen as a function of population size; to the extent the former are, ecological problems will tend to be linked instead to cultural orientations and patterns of economic and social organization. (Some other factors, notably technological capacity, would presumably be stressed by both sides.)

Biologists can readily envision population collapses, drawing analogies with other species (taken over a vast temporal span) and making use of a superior empirical knowledge of natural systems like the carbon cycle. Social scientists are more impressed by the rarity of collapses in human history and by the comparatively small effect on long-run population growth of what many would regard as extreme mortality conditions, like those associated with devastating civil wars. Whether continued population growth is moving human populations to qualitatively new terrain, where such lessons may count for less, is an issue very much in contention.

Perhaps the most important instance of drawing on different knowledge bases lies in systems theory itself. The profound changes in thinking about issues such as stability, continuity and predictability that have come from recent advances in understanding nonlinear systems behavior are highly relevant to both population ecology and theories about social change. Probably because they are more mathematized, population biology and ecology have taken greater cognizance of these developments; in the social sciences their effect as yet is fitful. Human demography, dealing with solidly overlapping generations and behavioral inertia, is seemingly as secure as ever in its understanding of its own system dynamics and is blithely unconcerned with possible—indeed, in the new dispensation, inevitable—surprises coming from the social and natural environment. It is likely that this is a simple lag effect that will be corrected over time. Robert May (1976) puts faith in better teaching, to refashion our intuitions about system dynamics, enriching them "by seeing the wild things that simple nonlinear equations can do." Conceivably, however, the problem is more deeply rooted in social scientific discourse—a possibility I consider below in discussing "habits of mind."

A second source of divergent views on demographic futures has to do with ethical and value premises. A broad range of value assumptions enters virtually any discussion of population and environment and often lies at the crux of the argument. In some cases, the matter may be important but the great majority of persons who consider it arrive at the same conclusion. This is the case, for example, for the argument that a per capita welfare criterion is inferior to one in which per capita welfare is weighted by population size. The latter, the so-called Benthamite criterion alluded to earlier, favors a much larger population. Despite a defence that some philosophers and economists have found persuasive, the Benthamite position is widely rejected. Parfit (1984) describes it as the "repugnant conclusion."

A case in which no such consensus has been reached is the issue, also mentioned above, of stewardship of the environment. From a biocentric standpoint, human appropriation of most of the biosphere is a deplorable thought. It is nearly equally so when our values are firmly anthropocentric, provided they incorporate a high weighting for wilderness and biodiversity.

In a well-known passage in J. S. Mill's *Principles*, the prefigured world to be avoided is one "with nothing left to the spontaneous activity of nature; with every rood of land brought into cultivation, which is capable of growing food for human beings; every flowery waste or natural pasture ploughed up, all quadrupeds or birds which are not domesticated for man's use exterminated as his rivals for food…" (Mill [1848] 1965). Considered in isolation, most people would doubtless support preservation of Mill's flowery wastes. The discriminating issue rather concerns appropriation at the margin: is this a matter for a quasi-constitutional decision or one for cost-benefit analysis? By and large, ecologists would opt for the strict limitation; social scientists—at least the more economically-oriented among them—would be found willing to trade. Each group can and does find ethical premises that justify its respective stance.

Preferences, of course, change over time and with satiety, and they adjust to the realities of what is achievable. "Wilderness," for example, is a comparatively recent object of desire in Western thought: Passmore (1974) describes the horror that wilderness used to evoke before the romantic movement transformed its image. If there were a clear time progression in preference systems toward conservationist ideals and increased respect for other species, as many people would argue, then a strong case could be made that policy judgments should take account of that trend and anticipate the emerging, more enlightened values. A constitutional rather than marginalist stance toward conservation would find support. Environmental preferences, however, not only record that progression: they also adapt to realities. Regret is held within bounds. Europeans protest the destruction of the tropical rain forests; they do not press for the reestablishment of their own temperate forests destroyed in earlier centuries. Dubos (1980, 129) remarks that some of today's most admired landscapes are the products or byproducts of human activity: "We struggle to save these human creations, forgetting that all of them represent areas deforested, swamps drained, hillsides gouged of their stones and sand."

Values are about means as well as ends. On the sensitive issue of anti-natalist policy in high-fertility societies, for example, strong environmentalist views sometimes translate into a clear-cut rejection of voluntarism, challenging the elevation of reproductive freedom to the status of a human right. It is argued, implicitly and occasionally explicitly, that birth controllers should be empowered to control births, a very different role from the one they are in practice granted—consisting of advertising the supposed benefits of small families and subsidizing the use of contraception. A similar harshness on the side of mortality yields the propositions of "lifeboat ethics" and "triage."

A third contributory explanation to the contrasting views of the demographic future may lie in "habits of mind." It is not always easy to separate what is entailed here from differences in knowledge and values, but the distinction is, I think, well worth making. In a well-known paper by Tversky and Kahneman (1981), judgment is held to be contingent on how the problem in question is "framed"—in turn, dependent on cues, habit, or experience. The actor does not see the latter as matters for deliberation and would usually not even be aware of them. In consequence, conflicting judgments may often trace back to differences in framing. It is plausible that disciplinary acculturation provides a powerful source of such framing cues.

In a more radical challenge to the rationality of cognitive processes, but broadly in the same tradition, Margolis (1987) proposes a theory of scientific theory-change in terms of "cognitive repertoires" of explanatory patterns and their relationships to cues. The objective is to account for "cognitive anomalies"—illusions of interpretation and judgment—that turn out to be independent of the intelligence, skill or vested interest of the persons making them. Such anomalies can encompass, at one extreme, simple and stylized Wason-type experiments in psychology and, at the other extreme, major transitions in world views such as the Copernican or Darwinian revolutions.

In many instances where anomalies arise, there is a "correct" choice—one that can be logically or experimentally demonstrated and, if the matter is sufficiently salient, that eventually dominates and excludes the other(s). This would model the now-standard Kuhnian view of scientific development. The course is onward and upward—though seen at close quarters, bumpy and replete with blind alleys. The same conceptual apparatus can also apply to situations where the alternatives are not theories as such but ways of apprehending some problem, and where the question of correctness in a scientific sense may not arise. In these situations there is the possibility for interplay of what Margolis refers to as "rival gestalts." Alternative conceptualizations of population-environment relations are clearly candidates for such an interplay.

Large intellectual differences may thus sit on a comparatively narrow substantive base. Over time, however, these differences acquire their separate evidentiary supports and, if they have significance in some public debate, garner political allies. Margolis (1987, 298–99) again:

> Facility at seeing things both ways is a cognitive burden [in a context of social controversy], where what is immediately useful is to be able to quickly see how a piece of information or an argument fits into the view to which you are committed, and to quickly see something wrong with arguments or information claims that do not fit that view.

Consequently, political polarization stimulates cognitive polarization, and in particular encourages the tendency for one view or the other to become dominant (easily seen, comfortably worked with). This is not, of course, a claim about what we consciously choose to do but about what we can be observed to do. Naturally, we find it harder to notice that tendency in our friends than in our adversaries, and most difficult of all to notice it in ourselves.

Convergent Futures

Simplify the demographic future by reducing it to an index: the world population total in 2100 rounded to the nearest billion. Assign a subjective probability to each number from zero up and consider the resulting distribution. It is not unlikely that persons who thought themselves far apart in their views on population matters, if they were separately to undertake such an exercise, would arrive at distributions that were not greatly at variance. Disagreement may thus be less over the probability to be attached to a demographic fall-back or collapse (fairly low), but rather to what should be inferred from an assessment that probability is some number appreciably greater than zero. Similarly, the probability of population reaching the neighborhood of the UN's high-variant projection for 2100 of 19 billion (again, fairly low) may be in little contention compared to the implications of such an outcome for humanity and for the natural world. If borne out, that would be progress toward a reconciliation.

A further desirable step in such a process would be to lessen the mutual disdain that characterizes the exchanges among some of the principal protagonists in the debate. Paul Ehrlich and Garrett Hardin, biologists of note as well as publicists of human overpopulation, are largely ignored by social scientists on that issue except as providing material for meta-study—say, for research on the rhetoric of the population problem or on fashions in crisis-mongering. (Hardin on the commons, in contrast, is established fare in several social science disciplines.) The disdain is reciprocated: economists are pilloried as willfully blind, demographers as wholly inconsequential. "Put together a list of the trained demographers who now occupy the most prestigious academic positions," Hardin (1988, 3) remarks, "then ask 'What deep insights into population problems have these men contributed?' I'm afraid one would have to answer, 'None.'"

This is not the place to assess demography's achievements and faults—I have written elsewhere on the subject (McNicoll 1992). It is, however, worth observing that an evident consequence of demography's self-identification as a social science and, by and large, its abjuring of ecological content has been to hobble it in any role it might aspire to in bridging the gap that I have been discussing in this essay. One way or another, with a bang or a whimper, a new demographic regime will emerge over the coming decades. It may look little changed from what we have now, merely a different allocation among already existing sub-regimes. (Recall Cassen's (1978) comment, I am sure an accurate perception, in reference to "the innumerable processes of adjustment" to population growth in India: "To a very considerable extent one can answer the question, what does the future hold for India, with the observation that the future has already arrived.") There is no great honor—though there would be a great deal of worthy investigatory work to be undertaken—in explaining in retrospect what happened demographically, and why. There should be, if not honor, at least a proper satisfaction with disciplinary duty done, in knowing that the span of alternative *futures* given serious consideration, both in analysis and in policy discussion, covered what, in fact, did happen.

References

Ashton, B., et al. "Famine in China, 1958–61." *Population and Development Review* 10 (1984):613–45.

Avery, D. "US Farm Dilemma: the Global Bad News is Wrong." *Science* 230 (1985):408–12.

Bos, E., P. W. Stephens, and M. T. Vu. *World Population Projections, 1992–93 Edition.* Baltimore: Johns Hopkins University Press for the World Bank, 1992.

Boserup, E. *The Conditions of Agricultural Growth.* Chicago: Aldine, 1965.

Bowden, M. J., et al. "The Effect of Climate Fluctuations on Human Populations: Two Hypotheses." In *Climate and History: Studies in Past Climates and Their Impact on Man,* T. M. Wigley, M. J. Ingram, and G. Farmer, eds. Cambridge: Cambridge University Press, 1981.

Cassen, R. *India: Population, Economy, Society.* New York: Holmes and Meier, 1978.

Cassen, R., and T. Dyson. "New Population Projections for India." *Population and Development Review* 2 (1976):101–36.

Cline, W. *The Economics of Global Warming.* Washington, DC: Institute for International Economics, 1992.

Conquest, R. *The Harvest of Sorrow: Soviet Collectivization and the Terror-Famine.* New York: Oxford University Press, 1986.

Crosson, P. "Agricultural Development: Looking to the Future." In *Sustainable Development of the Biosphere*, W. C. Clark and R. E. Munn, eds. Cambridge: Cambridge University Press, 1986.

Demeny, P. "Demography and the Limits to Growth." In *Population and Resources in Western Intellectual Traditions,* M. S. Teitelbaum and J. M. Winter, eds. Supplement to Population and Development Review 14, 1988.

Dubos, R. *The Wooing of Earth*. London: Athlone Press, 1980.

Feshbach, M., and A. Friendly. *Ecocide in the USSR*. New York: Basic Books, 1992.

Flöhn, H. "Singular Events and Catastrophes Now and in Climatic History." *Naturwissenschaften* 73 (1986):136–49.

Gilland, B. *The Next Seventy Years: Population, Food and Resources*. Tunbridge Wells, U. K.: Abacus Press, 1979.

_____ . "Considerations on World Population and Food Supply." Population and Development Review 9 (1983):203–11.

_____ . "Review of G. M. Higgins, et al. Potential Population Supporting Capacities of Lands in the Developing World." Population and Development Review 10 (1984):733–35.

Hardin, G. "Cassandra's Role in the Population Wrangle." In *The Cassandra Conference: Resources and the Human Predicament*, P. R. Ehrlich and J. P. Holdren, eds. College Station, Texas: Texas A & M University Press, 1988.

Higgins, G. M., et al. *Potential Population Supporting Capacities of Lands in the Developing World*. Rome: Food and Agriculture Organization, 1982.

Holling, C. S. "The Resilience of Terrestrial Ecosystems: Local Surprise and Global Change. In *Sustainable Development of the Biosphere*, W. C. Clark and R. E. Munn, eds. Cambridge: Cambridge University Press, 1986.

Hoyle, F. "A Contradiction in the Argument of Malthus." Reprinted in *Population and Development Review* 12 ([1963] 1986):547–62.

Jones, E. L. *Growth Recurring: Economic Change in World History*. Oxford: Clarendon Press, 1988.

Lamb, H. H. "Volcanoes and Climate: An Updated Assessment. In *Weather, Climate and Human Affairs,* H. H. Lamb, ed. London: Routledge, 1988.

Lederberg, J. "Medical Science, Infectious Disease, and the Unity of Mankind." *Journal of the American Medical Association* 260 (1988):684–85.

Margolis, H. *Patterns, Thinking, and Cognition: A Theory of Judgment*. Chicago: University of Chicago Press, 1987.

May, R. M. "Simple Mathematical Models with Very Complicated Dynamics." *Nature* 261 (1976):459–67.

McNicoll, G. "The Agenda of Population Studies: A Commentary and Complaint." *Population and Development Review* 18 (1992):399–420.

Meadows, D., et al. *Dynamics of Growth in a Finite World.* Cambridge, Mass.: Wright-Allen, 1973.

Mill, J. S. *Principles of Political Economy.* Toronto: University of Toronto Press, [1848] 1965.

Parfit, D. *Reasons and Persons.* Oxford: Clarendon Press, 1984.

Passmore, J. *Man's Responsibility for Nature.* London: Duckworth, 1974.

Peng, X. "Demographic Consequences of the Great Leap Forward in China's Provinces. *Population and Development Review* 13 (1987):639–70.

Revelle, R. "Food and Population." *Scientific American* 231 no.3 (1974): 160–70.

_____. "The Resources Available for Agriculture. Scientific American 233 no.3 (1976):165–78.

_____. "Let the Waters Bring Forth Abundantly." In Arid Zone Development: Potentialities and Problems, Y. Mundlak and S. F. Singer, eds. Cambridge, Mass.: Ballinger, 1977.

Richardson, L. F. *Statistics of Deadly Quarrels.* Pittsburgh: Boxwood Press, 1960.

Schelling, T. C. "Implications for Welfare and Policy." In Changing Climate, W. A. Nierenberg, et al, eds. Washington, DC: National Academy Press, 1984.

_____. "Some Economics of Global Warming." American Economic Review 82 (1992):1–14.

Sen, A. K. *Poverty and Famines: An Essay on Entitlement and Deprivation.* Oxford: Clarendon Press, 1981.

Simon, J. L. *The Ultimate Resource.* Princeton: Princeton University Press, 1981.

Smil, V. "Population Growth and Nitrogen: An Exploration of a Critical Existential Link." *Population and Development Review* 17 (1991):569–601.

Steel, D. "Our Asteroid-pelted Planet." *Nature* 354 (1991):265–67.

Tversky, A., and D. Kahneman. "The Framing of Decisions and the Psychology of Choice." *Science* 211 (1981):453–58.

United Nations. *Long-Range World Population Projections: Two Centuries of Population Growth 1950–2150.* New York: United Nations, 1992.

Vitousek, P. M., et al. "Human Appropriation of the Products of Photosynthesis." *BioScience* 36 (1986):368–73.

Watkins, S. C., and J. Menken. "Famines in Historical Perspective." *Population and Development Review* 11 (1985):647–75.

World Health Organization. *Effects of Nuclear War on Health and Health Services, Geneva.* Excerpted in *Population and Development Review* 9 (1983):562–68.

Wright, Q. *A Study of War.* Chicago: University of Chicago Press, 1942.

11

Population, Constraint and Adaptation: A Historical Outlook

By Massimo Livi-Bacci

The first time I met Roger Revelle was at a meeting organized by Daedalus (actually, by Revelle and Stephen Graubard) on Population and History at the Bellagio Center in the late 1960s. Ansley J. Coale, Paul Demeny, Etienne Van der Walle, Bernard Slicher van Bath, John Hajnal and others were present. The meeting lasted one week in the splendid frame of the Villa Serbelloni. The discussion was wide-ranging and very informative for a young person such as I. For the first time, I fully appreciated the potentiality of linking such different themes as environment, resources, economy and population variables in a historical perspective. The fact that two non-demographers were actively coordinating and stimulating the conference was very productive; Revelle, a biologist, and Graubard, a historian and political scientist, forced demographers to venture outside of their own area. I met Revelle in the following years on a number of occasions, and I have always been grateful for the opportunity he gave me to approach my research interests from a different perspective.

Introduction

"It may be safely asserted…that population, when unchecked, increases in a geometrical progression of such a nature as to double itself every twenty-five years….Practically, it would sometimes be slower, sometimes faster." (Malthus [1830] 1970). Demographers would readily subscribe to this statement: a doubling time of 25 years means an annual growth rate of close to 3 percent—the common lot of developing societies in the second part of this century—"sometimes slower, sometimes faster" depending on local conditions, but faster than in any past historical period we know about.

Malthus was referring to the United States—a country where there was an abundance of equitably distributed land, which produced for open markets, and which had a high demand for labor and high labor productivity, and a minimum of positive checks and no need for the preventive ones. In Europe, where land was less abundant, its distribution often greatly inequitable, with a denser population and scarcer "necessities," the doubling time was much lower even in favorable times free from catastrophes. Historical population growth is shaped by the interaction of the "natural power of mankind to increase" (Malthus [1830] 1970) and the checks imposed for want of space, land and nourishment.

This paper is one further attempt to reformulate Malthus' theory with the additional knowledge accumulated during the last two centuries. I will start by defining growth and its dimensions, and examine the way in which different populations fit in the space so defined. I will call this "strategic space" meaning that populations may move (by choice or constraint) in this space, so determining their historical trajectory. Next, I will briefly touch upon the complexity of the factors of constraint—the obstacles or check to population growth. Then, I will discuss the mode of operation of those mechanisms of adaptation and choice that make it possible for populations to react or adapt to the forces of constraint. Finally, I will argue that those populations with maximum flexibility in the face of the forces of constraint are likely to be more successful, and I will discuss the implications of this assertion.

The Strategic Space of Growth

Fertility and mortality, acting in tandem, impose objective limits on growth rates of human populations. In a steady state and with some simplification we can express the rate of growth as a function of the number of children per woman (total fertility rate or *TFR*) and life expectancy at birth $e(o)$ (Livi-Bacci 1992). Figure 1 represents a space defined by fertility *(TFR)* and mortality *(e)*. Each curve, known as an "isogrowth curve," is the locus of those points that

FIGURE I

Relations Between the Average Number of Children Per Woman (TFR) and Life Expectancy (e_0) in Historical and Present-Day Populations

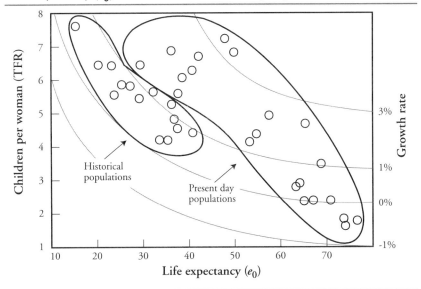

FIGURE 2

Relations Between TFR and e_0 in Historical Populations

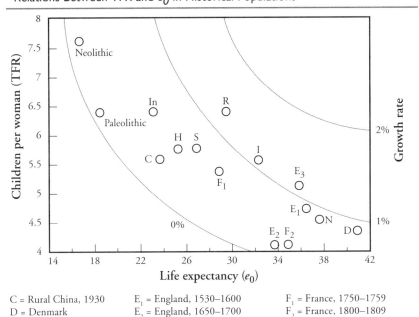

C = Rural China, 1930	E_1 = England, 1530–1600	F_1 = France, 1750–1759
D = Denmark	E_2 = England, 1650–1700	F_2 = France, 1800–1809
I = Italy, 1862–1867	E_3 = England, 1750–1800	N = Norway, 1780
S = Spain, 1787	H = Hungary, 1830	R = Russia, 1897
In = India, 1900		

FIGURE 3

Relation Between TFR and e_0 in Present-Day Populations

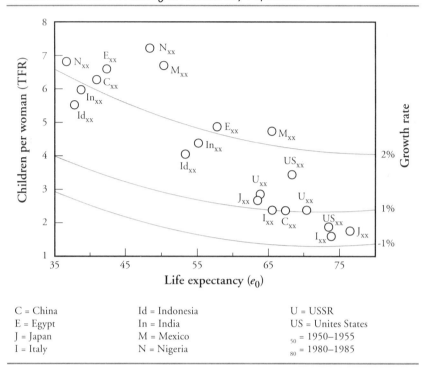

C = China
E = Egypt
J = Japan
I = Italy

Id = Indonesia
In = India
M = Mexico
N = Nigeria

U = USSR
US = Unites States
$_{50}$ = 1950–1955
$_{80}$ = 1980–1985

combine life expectancy (the abscissa) and number of children per woman (the ordinate) to give the same rate of growth r. The graphs contain points which represent points corresponding to historical and contemporary populations. For the former, life expectancy is neither below 15, as this would be incompatible with the continued survival of the population, nor above 45, as no historical population achieved a higher figure. For similar reasons the number of children per woman falls between eight (almost never exceeded in normally constituted populations) and four (historical populations did not practice birth control in a modern sense). In contemporary populations the upper limit of life expectancy is above 80 and the lower limit of fertility approaches one child per woman in some western societies. In Figure 1, historical populations are distinguished from present-day populations (second half of our century); Figures 2 and 3 show these in more detail.

Most historical populations (Figure 2) are situated within a relatively narrow space of growth comprised between the isocurves corresponding to 0 and 1 percent. Within this narrow band, however, fertility and mortality combinations vary widely. Denmark, at the end of the eighteenth century, and India, at the beginning of the twentieth, for example, have similar growth rates. These, however, are achieved at distant points in the strategic space described: Denmark combines high life expectancy (around 40 years) and a small number of children (just over four), while in India, low life expectancy (25 years) is paired with many children (just under seven). Figure 3 represents some of the most populous countries since 1950. The strategic space utilized, previously restricted to a narrow band, has expanded dramatically. Life expectancy has been raised from 20 to 40 years, common in historical populations, to above 80, and the diffusion of birth control has pushed down the lower limit of fertility to about one child per woman. In this much expanded space the populations represented vary between a potential maximum growth of 4 percent a year (many developing countries have experienced growth rates well above 3 percent during the last decades) and a minimum of –1 percent (the rate of growth of Germany, Italy and Spain should their current fertility remain unchanged for a long period). The two situations depicted in Figures 2 and 3 differ not only in the dimensions of the strategic space they occupy, but also, and especially, in their duration. The first of the two figures represents potential historical situations of long duration, while the second is certainly unstable and destined to change rapidly since it implies patterns of growth that cannot be sustained.

The strategic space may be further decomposed: the fertility dimension *(TFR)* is a function of the proportion of the population in fecund age living in stable union (marriage) and of fertility in marriage. The former measure— or proportion married, in societies where marriage is the "locus" of fertility— may itself be decomposed into age at marriage and proportion ultimately getting married. On the other hand, marital fertility is an inverse function of the intervals between births. The length of birth intervals, even in societies not controlling fertility, varies widely as a consequence of duration of breastfeeding, coital frequency, and intrauterine mortality. Also the expectation of life—the other dimension of the strategic space of Figure 1—can be decomposed into many factors, although these cannot be considered elements of a conscious strategy. We all agree on the existence of an inborn instinct of survival; however individual, family or societal arrangements are very relevant for infant and child survival. There is an element of conscious strategy in this—investment in children takes the form of breastfeeding; sheltering the infant from cold or heat; isolating them from a dangerous environment; feeding them after weaning, and so on. Investment in infants means competition in the allocation of resources among family members, and this may explain why there is no direct correlation (and often an inverse correlation) between mortality of the very young and mortality of adults. It follows that for an identical expectation of life at birth there are different combinations of infant-child and young-adult mortality.

Historical populations appear to occupy a "narrow" strategic space, with rates of growth mainly comprised between 0 and 1 percent. This strategic space is "narrow" only when compared with the exceptional modern period: indeed, small differences in the rate of growth have large implications for absolute numbers over generations or centuries. The European population has increased at rates of –0.3 percent between 1300 and 1400; 0.25 percent during the 15th and 0.28 percent during the 16th centuries; and 0.06 percent during the 17th and 0.43 percent in the 18th centuries (Biraben 1979). If the 16th century rate had been sustained for an additional 150 years, the European population would have doubled—an impressive achievement in a society with limited land and relatively fixed technology—during this same time span, with the growth prevailing in the 17th century, the population would have grown a modest 18 percent. Thus, a modest variation in the strategic space has very relevant implications for long term growth. Larger variations in growth rates can be found across populations, and the results are more dramatic among smaller groups.

Understanding Growth

There is a strategic space—extremely simplified in Figure 1—in which populations are allowed to move about: in other words, populations have the capacity to grow (and to change their rate of growth) within well-specified limits. Although bound in the past within a narrow strategic space, the combinations of fertility and mortality could lead a population to decline and disappearance or to rapid growth. For demography, fundamental questions are why certain populations have been more successful than others; why a given population has a certain dimension and is not much smaller or much larger; why the world population is approaching 6 billion people and does not number 6 million as in the late Paleolithic era or 60 billion as some say our planet could support. It is easy to agree that, in the long run, millennia or centuries being the unit of measure, there must be some harmonic development of population and resources. A good depiction of this conviction is provided by a well known curve of population size—time and numbers on a logarithmic scale from the origin of mankind to the present day. The curve is composed of three segments intersecting in two revolutionary phases, each segment describing, initially, a high growth rate slowly decelerating into a terminal phase of quasi-stagnation (Deevey 1960). The first phase terminates in the late Paleolithic, when our hunting and gathering predecessors saturated the carrying capacity of the planet. The second phase initiates with the Neolithic revolution, some 10,000 years or so ago, and terminates with the onset of the industrial revolution. In this phase the limiting factor was land, for the production of food and energy, and the slow growth of agricultural productivity. The third phase initiates with the industrial revolution and the invention of inanimate converters for the exploitation of new sources of energy (coal and steam engine) which made it possible to multiply the supply of energy that had limited growth in the preceding phase. The third phase, therefore, began some two centuries ago, but nobody knows or can predict when it will come to a close.

This simple interpretation of past trends is probably acceptable to many. And, indeed, it is difficult to accept the idea that the society of the Paleolithic could have supported more than a few million people hunting deer and boars, and picking fruits and berries, or that the agricultural techniques available two centuries ago, from the more advanced of some areas of Asia and Europe to the most primitive ones, could have fed the more than the one billion people currently estimated to have inhabited the world at the beginning of the 19th century. On this, perhaps with some hesitation, we can all agree, as we do agree with Adam Smith's splendid tautology that "every species of animal naturally multiplies in proportion to the means of their subsistence, and no species can ever multiply beyond it" (Smith 1964). This is so because the interpretation of demographic growth I have stated above relates to millennia and is ex post facto; in the very long run, material resources and population seem to proceed in harmony. But if instead of millennia or centuries, we use as our unit of measure shorter durations such as the life span, the length of a generation, or decades or years, then the harmonic view proposed above vanishes. Sussmilch's "divine order" gives way to the wild "struggle for existence;" populations grow and prosper at the expense of others; some retreat and vanish, others maintain precarious equilibria.

Constraint and Adaptation

We may conceive of demographic growth as taking place within two great systems of forces, those of constraint and those of adaptation. Among the forces of constraint, we can list the limitation of land and consequently of food and energy—resources that satisfy basic needs. Carlo Cipolla observed that "the fact that the main sources of energy other than man's muscular work remained basically plants and animals must have set a limit to the possible expansion of the energy supply in any given agricultural society of the past. The limiting factor in this regard is ultimately the supply of land" (Cipolla 1962). In preindustrial Europe, populations seem to have frequently approached the limits allowed by available land and existing technology. These limits may be expressed by the per capita availability of energy and, again following Cipolla, must have been below 15,000 calories, or perhaps even 10,000 per day, the majority of which were dedicated to nutrition and heating. The availability of space also acts as a constraint not only for the obvious limitations it puts on settlement but also in relation to the risk of onset and spreading of infectious diseases. In the Neolithic period, sedentarization created the conditions necessary for the onset, spread and survival of parasites and infectious diseases which were unknown or rare among mobile or low-density populations. Higher demographic concentration acts as a "reservoir" for pathogens, which remain in a latent state awaiting an opportune moment to resurface. The spread of diseases transmitted by physical contact is favored by increased density, and this density in turn increases the contamination of soil and water, facilitating reinfection. With settlement, many animals, domesticated or not, come to occupy a stable place in the human ecological niche, raising the possibility of infection from animal pathogens and increasing the incidence of parasitism. Agricultural technology may also have been responsible for the spread of certain diseases such as malaria, which benefited from irrigation and the artificial creation of pools of stagnant water. Growth of cities and urbanization in general have increased some of these problems. Environmental characteristics produce other constraints on population, independently from the limits they impose on settlement, on land and food availability, etc. An example is climate, and the checks that extreme heat or cold place on a population and its survival. Epidemics have been another important force of constraint; their diffusion and resurgence is only partly dependent on space and density and, beyond certain levels, independent from them. Plague affected the whole of Europe independently from density and patterns of settlement.

Limitations of space and land; poverty of the food and of the energy supply; the hostility of environment and climate; the attacks of infectious diseases and of epidemics are examples of forces of constraint that have checked population growth in various ways. These factors of constraint are relatively fixed in time. Mankind can only partially modify and control these forces and then only over long periods of time, usually exceeding the length of a generation or the life span. A population can learn to protect itself from epidemic disease, increase the availability of food or energy in spite of land limitation or lessen the negative effects of climate; but the implementation of measures to prevent contagion, the improvement of techniques of cultivation for increasing energy and food availability, or the adoption of dwellings better suited for harsh climates are all processes that require long stretches of time. In the short or medium term (but, often also in the long) population must adapt to the forces of constraint.

It is obvious that in the very long run, human progress—accumulation of knowledge and of material wealth—is the response to the action of constraint. But the adaptation is another way to minimize, to delay, or to avoid the adverse effects of the forces of constraint; it ensures the flexibility that a population needs in order to adjust its size, growth and structure to the action of those forces. This is not to say, as is often rashly asserted, that populations are provided with providential regulating mechanisms (homeostasis) that maintain size and growth within dimensions compatible with available resources. There is nothing automatic or necessary in this process of adjustment, as is shown by the fact that many populations have disappeared and others have grown to such a degree that equilibrium could not be restored without much pain and suffering.

Demographically relevant adaptation is partly biological in nature and partly genuinely behavioral, the first largely automatic and unrelated to choice, the second mainly determined by it. Neither of them, as demonstrated below, operates in a fixed manner.

Adaptation: Biological Aspects

The fact that biological processes are not under the control of the individual has often discouraged demographers from taking them into account. We tend to believe that these processes are more or less invariant in the history of mankind and that we react, biologically, to constraints like disease and want in an unchanging way. I shall argue that this view is not correct. The interplay between host and infectious agents is not completely understood but the action of the major causes of death tends to change over time. Typical is the case of virgin soil populations—like those of America and the Pacific at the time of contact with the European— lack of immunity led to extremely high mortality from diseases that were common in Europe and Asia but new to them, leading these populations to precipitous decline and, in some cases, and for smaller communities and groups, to extinction. But exposure created immunity, often passed from one generation to the next; defenses were progressively built up, mortality declined, and population size recovered.

The demographic collapse of indigenous populations as a result of contact with Europeans is a widespread and well-known phenomenon throughout America and Oceania. The Indios of Santo Domingo and of Cuba became extinct during the 16th century; a similar fate was encountered by the indigenous population of the Tierra del Fuego in the last two centuries; Darwin relates the extinction of the Tasmanians during the 19th century.

Larger populations, such as those in North America, Meso-America or the Andes, the Maori of New Zealand or the Aborigines of Australia, suffered a precipitous decline lasting several generations before they could initiate a recovery. Demographic losses were due to a variety of lethal diseases. In addition to smallpox and typhus—scourges in the Old World, as well as the new—tuberculosis, measles, influenza and chicken-pox all left their mark (Livi-Bacci 1992).

This process of adaptation evident in virgin soil populations is evident also in other populations where the virulence of important infectious diseases, and therefore the related mortality, appears to have changed over time: syphilis, malaria, tuberculosis, measles and influenza are all cases in point. Selection in the human population and, much more importantly, mutations and selection in the pathogens, account for the changing importance, over time, of given diseases. It has been argued that if the plague had remained constantly present and attacked a large portion of the new generations as they were formed, it might gradually have assumed an endemic, sporadic form, with relatively low mortality.

We may believe that mortality is, nowadays, under control, or at least in the process of being brought under control. And it is so, if we look at the startling progress of the expectation of life things achieved almost everywhere. But "nothing in the world of living things is permanently fixed…on purely biological grounds, therefore, it is entirely logical to suppose that infectious diseases are constantly changing; new ones are in the process of developing and old ones modified or disappearing…it would be surprising therefore if new forms of parasitism—that is, infection—did not constantly arise and if, among forms, the modifications in the mutual adjustments of parasites and hosts had not taken place during the centuries for which we have records" (Zinsser [1935] 1971). Prophetic words, indeed. The appearance, diffusion, lethality, and changing nature of AIDS is the terrible proof of the truth of the prophecy.

Another area in which biological reactions are extremely important is that of nutrition, an important constraint of survival. The human body has a remarkable ability to adapt to nutritional stress through a number of mechanisms which vary according to the character and nature of the hardship or scarcity suffered (Livi-Bacci 1991). Humans, when coping with short-term nutritional stress, react by trying to reestablish an energetic equilibrium at a lower level, by burning fat reserves, and expending less energy as a result of lower body weight, for a given level of physical activity. It is also noted that the basic metabolism seems to change more than proportionally in relation to weight, due to a more efficient consumption of the energy available. In more serious cases, reduced physical activity can further lower the threshold of caloric equilibrium.

This capacity to adapt over short periods is innate in the human species, forced to survive hundreds of thousands of years in unstable environments subject to sharp climatic variations. Mankind has developed a short-term strategy for dealing with temporary nutritional stress and minimizing the harm that can be done by the vagaries of the weather, a poor harvest, or even a whole year of failed harvests. It is a strategy whose time-span may be a month, a season or perhaps even a year. This flexibility has allowed the hunting and gathering Bushmen of the Kalahari to survive the dry season on 1,000 calories less per day than they can expect to consume in the good season. For the developing world, such variations in food availability are still frequent, as they were in many populations in the past.

There also exists a medium-term adaptability, which comes into operation in cases of prolonged nutritional stress and enables a population to develop a new equilibrium between food resources and reduced levels of consumption. This second line of defense entails reduced body growth in childhood and adolescence. Where stress is moderate, growth is slowed down while keeping balance between weight and height. People whose growth is retarded in this fashion are not necessarily less efficient than others with normal growth. The deviation of actual growth from the growth genetically possible under optimal food conditions (ignoring the negative influence of infections) can theoretically be split into two parts. One part is not connected with a heightened risk of infection and death, and is thus, successfully adaptive; but over a certain threshold, which is difficult to determine, the risk of death increases, and this represents the nonadaptive part of deviation. In the very long term this nonadaptive lack of growth may also become genetically selective: but if so, its effects would not be discernible within the time-span familiar to demographers. However, in the medium to long term (a time-scale of decades or generations or at most centuries) a slowing down of skeletal body growth such as occurred in some European populations during the latter part of the eighteenth century would seem an efficient response to diminished food availability. This is a mechanism which tends to mitigate the negative effects that a slow reduction of food resources, provided these remain above a certain threshold, might otherwise have on survival. Thus the human phenotype may assume body size compatible with the limits imposed by its environment and growth is optimized relative to the resources available. If stunting—as it occurs in some contemporary poor countries—limits adult male body weight to 60 kgs instead of 70 kgs, (with improved metabolic efficiently) this would imply a 20 to 25 percent reduction in energy required for a given level of physical activity, a highly efficient strategy for survival. Obviously, if an equilibrium is not reached, physical activity will have to be reduced and this harms the efficiency of the individual and eventually leads to social regression. But this need not happen: if the equilibrium is reached, reduced body size does not constitute a disadvantage and in some environments appears to constitute an actual advantage.

Above a certain threshold, prolonged nutritional stress does not allow the adaptive process to avert the damage reflected in increased mortality. One can, however, imagine that a long history of want and deprivation might exercise a selective pressure in favor of those organisms which make the most efficient use of meager food supplies. Authoritative opinion maintains that prolonged under-nourishment may result in the negative selection of children with greater growth potential, who are more likely to perish. The small stature characterizing populations which have suffered from a long history of food shortage may have a genetic explanation. This would account, for instance, for the lower stature of Meso-American populations. This mechanism, however, acts only over the very long term; it is not adaptive in the sense in which we have so far used the expression because it is connected with higher mortality.

The last type of adaptability is that which enables populations to absorb gradual changes in the composition of diets without reducing their capacity for survival. One thinks, for example, of the capacity of the Eskimos to survive on a diet nine-tenths of whose caloric content is derived from animal foods; of the ability of certain African populations who lack salt to limit its loss in sweat and urine; and of the vast range of diets consumed by contemporary populations, from those exclusively vegetarian to those overrich in animal nutrients. We can presume that over the long term, populations have had no problems in adapting even to considerable changes of diet brought about by climatic changes or other factors. Damage occurs, however, in situations of rapid change when biological or, above all social, adaptability does not have sufficient time to come into operation. For instance, the introduction and diffusion of maize, economically advantageous to peasant families, gave rise to pellagra, and sudden changes in diet due to migration have often caused other imbalances. In modern times, there are examples of rapid changes and dislocations—from rural to urban areas, from self-production to marketed food—that have lowered the ability to adapt to shortages or changes in diet. The connection between shortened breastfeeding, changes in diet, and infant mortality in many populations in Africa is a case in point: one wonders whether the precipitous abandoning of traditionally experienced models of nutrition leads to a decreased resistance to nutritional stress.

Social Dimensions of Adaptation

Social processes of adaptation, which can be defined as mechanisms of choice (among them Malthus' preventative checks) are much more familiar to demographers, and were referred to earlier in this paper, in the discussion of the strategic space of growth. It is common wisdom that mankind is, nowadays, much more in charge of its own numerical destiny than in the past: in other words, we are much more able than in the past to choose the patterns of demographic change. This is only partly true: indeed during the last two hundred years women and men have been learning how to control fertility. But mankind is now a much faster vehicle than it used to be, and needs a much stronger braking system than in the past.

I shall comment briefly on three main mechanisms of choice: the one regulating access to reproduction, or marriage; the one regulating fertility; and the choice of the place of settlement (migration, mobility). Marriage, as a mechanism of the regulation of population growth has been a powerful tool in many populations of the past. In western societies, delayed marriage (with mean age for females close to 25 years) and high rates of celibacy (up to one fourth or one fifth of a generation never marrying) were common. Changes in these two parameters would greatly affect the proportion of the population living in a fertile marriage at any given time and thus, indirectly regulate fertility. Regulation of marriage could operate in three different (but not independent) ways:

a. In the short term, as in response to fluctuations of agricultural production, marriage could be anticipated or delayed according to necessity. After a demographic crisis, losses would be minimized by the anticipation of marriage and remarriage of the widowed; a number of economic and social "niches" would become free because of exceptional mortality and would be filled with newly founded families. This process occurred after the major mortality crisis in the European populations.

b. In the medium to long run, changes in economic conditions (for instance an increase or a decline of real wages) would bring about changes of nuptiality which in turn would affect fertility and the rate of growth. The English demographic system from the 16th to the 18th century seems to have worked this way, keeping population from growing too quickly in economically difficult times or accelerating its growth during the periods of prosperity (Wrigley and Schofield 1981). In other countries the system has been less efficient and flexible than in England. The high growth of the population of Ireland during the 18th century and the first decades of the 19th century seems to have been propitiated by an increase of nuptiality, a consequence of the higher productivity of land where the potato was introduced as the main crop; but the system did not capture the signals of economic stress (that could have been translated into decreasing nuptiality) that led to the crisis of the 1840s (Connell 1950).

c. A change in nuptiality patterns could be produced by radical and relatively permanent modifications in living conditions and the social structure (Livi-Bacci 1992). The demographic success of the French pioneers in Canada was partly due to an age at marriage lower than in the areas of origin. In Japan, before the Tukogawa era, serfs were largely excluded from marriage, but the abolition of the restriction was probably an important factor of the dynamic 17th century population growth. In Ireland, the crisis of the 1840s brought about a radical change in nuptiality with very high age at marriage and high levels of celibacy.

In developing countries today, marriage is a much less powerful regulator, both because access to marriage is nearly universal, and because social norms prescribe, in many countries, an early marriage. Marriage, the typical check that Malthus deemed appropriate for prudent populations, while an efficient regulator of growth in high mortality, slow growing populations, is much less powerful in low mortality, fast-growing societies. In spite of signs of an increasing age at marriage in Asia and Latin America, changes in nuptiality are responsible for a relatively small part of the decline of the number of children per woman in recent times.

Regulation of fertility in the modern sense—implying conscious choice of timing and number of births—is a relatively recent process, initiated some two centuries ago in the west and still unknown in many poor societies. But in societies with natural fertility, fertility may vary considerably because of different durations of lactation, coital frequency, etc. Pretransitional regional levels of fertility in Europe showed considerable variation. Whether this variation was associated with an element of choice is debatable; many will agree, however, that variation of fertility represented an important factor of demographic flexibility. Fertility, like nuptiality, had a certain amount of flexibility evident in the short-term (economic fluctuations associated with fertility fluctuations); an increase of fertility has also been observed as a medium-term reaction to mortality crisis. In modern times, fertility control has become the true regulator of growth, increasing in strength almost everywhere with the notable exception of tropical Africa and a number of Asian societies. The increase in contraception, however, in some instances does not compensate for the decline in the duration of breastfeeding; this implies a shortening of birth intervals and an increase of fertility. In other words, the efficiency of fertility regulation may be, for periods more or less long, on the decline. Finally, with low mortality, demographic growth can be regulated mainly by the vigorous action of fertility control (the role of nuptiality being reduced as explained above). Its modern diffusion and impact is still gravely inadequate to "narrow" the strategic space of growth. Hence the paradox: fertility control as a factor of choice and of adaptation of growth to constraint has increased in modern times (diffusion of contraception and improved efficiency) while its power to regulate growth has declined.

A third traditional regulator is migration. World population patterns of settlement are the product of migration: from Asia to Europe, from Africa to America, from Central Asia to Southern Asia, etc. The Neolithic revolution implied colonization and settlement of new lands, and diffusion of agriculture took place with successive waves of migration. One wonders what the features of the industrial revolution would have been without the massive emigration from Europe to America, without the "colonization" (and I am using the word in the ordinal sense) by people, plants and animals of the temperate regions of the world. One pictures the history of mankind as marked by a continuous process of population redistribution through which, an economist would say, an optimal allocation of human resources is pursued. This is, of course, an exaggeration, since barriers of all types have always hindered this process of redistribution; invaders and settlers fought, borders were guarded and protected, intruders deported. But the process of redistribution, until this very century, has gone on, new territories have been opened and settled, different populations have mixed. We could say that migration and settlement, as instruments of choice, were potentially very efficient.

During this century the logic of the national state and of this substantial demographic integrity has rapidly and sometimes violently affirmed itself. Nowadays, migration is possible and acceptable only in the interest of the receiving state, as one possible factor for the regulation of the labor market. That the interests of the receiving state may partially coincide with the interests of the sending one is of secondary importance. The second World War and the subsequent rapid process of creation of new independent national states in Africa and Asia has in many instances imposed conventional borders, cutting across important minority ethnic groups or religious communities. These groups and communities were often forced to retreat to their original homelands, imposing a redistribution process of a "negative" sign.

The causes of the contemporary inversion of historical patterns are many. Probably the most important one is the gradual disappearance of open territories as a consequence of population growth, but the differentials in opportunities and standards of living between countries and areas of the world, together with the revolution in transportation, have also made impossible the continuation of past trends. Many states of the world are now able to close their frontiers as they wish. In many cases massive flows of "undocumented" migrants exist only because of the implicit connivance of important political or economic forces in the receiving country.

The progressive sclerosis of migration also affects, in some cases, internal movements for a variety of reasons, thus further restricting the scope of migration as an instrument for the optimal combination of population and resources.

Conclusion

Understanding long term population growth implies three major efforts. The first must clarify the nature of the forces of constraint and their impact on demographic variables: land, climate, pathologies and epidemics, that is the various dimensions of the environment. The second requires understanding the impact of human action in modifying constraints. In other words, the study of accumulation of knowledge and wealth. The third step consists in understanding the mode of operation of the forces of adaptation to constraint, in some case biological in nature and to a certain degree, automatic, in others the product of choice.

In this paper, I have argued that the forces of adaptation enable individuals and groups to react to those of constraint; that they ensure the flexibility essential for the survival or the well-being of a population: the greater the flexibility, the larger the options of a population in responding to constraints. A population where the forces of adaptation are weak or have no time to operate is like an engine with a thermostat out of order, a vehicle without brakes, a body insensible to pain. It can be argued that, in modern times, the forces of choice have increased their strength by the diffusion of fertility control. But the process is not linear: increasing density, the lack of open spaces, the strengthening of national borders have weakened migration (regulator of the geographical distribution of the population) and as a consequence, the pursuit of equilibrium, however precarious and unstable, between population and resources, has lost several "degrees of freedom." Biological adaptation may face new challenges: thousands of years have made mankind fit to withstand nutritional stress, or the variability of climate or the physical fatigue of hard work. But the changing environment and the changing rhythm of life produce other constraints that require the development of other types of adaptation and reactions. Mankind is a much faster vehicle than in historical times but its braking system has not been adequately reinforced.

REFERENCES

Biraben, J. N. "Essai Sur L'evolution du Nomre des Hommes." In *Population* 1 (1979).

Cipolla, C. M. *The Economic History of World Population*. Harmondsworth: Penguin Books, 1962.

Connell, K. H. *The Population of Ireland, 1750–1845*. Cambridge: Clarendon Press, 1950.

Deevey, E. S. "The Human Population." *Scientific American* September 1960.

Livi-Bacci, M. *Population and Nutrition*. Cambridge: Cambridge University Press, 1991.

_____. A Concise History of World Population. Cambridge, MA: Blackwell, 1992.

Malthus, T. R. *A Summary View of the Principle of Population*. Harmondsworth: Penguin Books, [1830] 1970.

Sanchez, A. N. *La Poblacion de America Latina desde los Tiempos Precolombinos hasta el 2000*. Madrid: Alianza Editorial, 1977.

Smith, A. *The Wealth of Nations,* vol. 1. London: J. M. Dent and Sons, 1964.

Wrigley, A. E., & R. S. Schofield. *The Population History of England, 1541–1871*. London: Arnold, 1981.

Zinnser, H. *Rats, Lice and History*. New York: Bantam, [1935] 1971.

APPENDICES

ROGER R. REVELLE

Director Emeritus of the
Scripps Institution of Oceanography
and
Professor Emeritus of
Science and Public Policy

1909–1991

Roger Randall Dougan Revelle, renowned oceanographer and founding father of UCSD, was one of this century's most eminent men of science. He won admiration both for his research and for his ability to promote and direct scientific inquiry and education innovation on a grand scale. By inspiring others with his own infectious zest for learning and by taking pains to identify talent and credit achievement, he brought out the best in those with whom he worked. Imaginative, energetic, forthright, and always ready to take up a worthy cause, he was also an exemplary citizen of his community, his country and the world.

Roger Revelle was born in Seattle March 7, 1909, to a family of Huguenot descent. While he was still a child, his mother came down with tuberculosis. To give her the benefit of a warmer climate, his father withdrew from the law practice he owned with his brothers and moved the family to Pasadena, where he became a school teacher and did legal work from a home office so that he could have more time to take care of her and the children. It was there that Roger received his schooling and first attracted notice: he was among the schoolboys who earned high scores in an intelligence test administered by the psychologist Lewis Terman and whose careers were monitored by Terman to vindicate the view that nature mattered more than nurture.

Because of the promise he showed in high school, Revelle was admitted to Pomona College at the age of sixteen. There he thought of becoming a journalist until, in one of his science courses, he fell under the spell of a charismatic teacher, Professor Alfred O. "Woody" Woodford, who became legendary for inspiring students to take up geology. After receiving his B.A. in 1929, he stayed at Pomona to continue working with Woodford for another year and then went on to pursue graduate studies in the subject at UC Berkeley.

In 1931, while still a graduate student, he was brought to the attention of T. Wayland Vaughan, Director of the Scripps Institution of Oceanography (then a remote field station), who invited him to join a project at SIO studying deep-sea mud. By coincidence, shortly after agreeing to go to SIO, he married Ellen Clark, whom he had been courting since her sophomore year at Scripps College and who is a grandniece of both Edward Willis Scripps and Ellen Browning Scripps, in honor of whose benefactions the Institution had been given the family's name.

Upon receiving his doctorate in 1936, Revelle spent a year at the Geophysical Institute in Bergen, Norway, working under the Arctic explorer, Bjorn Helland-Hansen, returning afterward to accept appointment at SIO as an instructor. He also joined the Naval Reserve, and took leave to enter active service when the U.S. declared war in 1941. Initially posted to the Navy's radio and sound laboratory in San Diego, where he felt his training as an oceanographer could not be put to proper use, he managed to persuade the higher-ups to reassign him to Washington in 1942 to serve as officer-in-charge of the oceanographic section of the Bureau of Ships. He entered active duty as a lieutenant, j.g., and retired at the end of the war with the rank of commander in the Naval Reserve.

While in Washington, he played a role in helping to organize the Office of Naval Research, which was to serve as the first conduit for the federal government's support of basic research in the universities and the model for the National Science Foundation. From 1946 to 1948, he headed ONR's Geophysics Branch. In this capacity, he was assigned to organize the oceanographic aspects of the atomic bomb tests, including Operation Crossroads, the first peacetime test at Bikini Atoll in 1946. The study he designed examined the marine biology of the atoll both before and after the test, providing data of major scientific importance and instilling in him a lifelong concern for averting nuclear war.

In 1948 Revelle returned to SIO, this time as a full professor. In rapid succession, he became associate director (1949–50), acting director (1950–51) and finally director of SIO in 1951, making good use of his contacts at ONR to obtain the funding needed to expand the Institution's activities.

Although Revelle's appointment as director did not meet with universal enthusiasm—one SIO researcher complained that he was too untidy to be trusted with administration, noting that he "just let everything pile up on his desk" and "was too easily diverted"—it was under his leadership that SIO began to achieve international standing and set the foundations of much of modern oceanography. He personally led the Mid-Pacific Expedition and the nine-month Capricorn Expedition. "Basically," he observed afterward, "what I did was to send the institution out to sea, to make it a worldwide institution instead of just a local California institution. The farthest we ever went before the war was the Gulf of California….By the time I left we had a Navy that ranked with that of Costa Rica and had sailed literally millions of miles everywhere in the world." Oceanographers, he liked to say, "are just sailors who use big words."

Big words, he could have added, with even bigger consequences. SIO's expeditions found evidence of magnetic reversals, mantle convections, the formation of manganese nodules and many varieties of hitherto unknown deep-sea life forms. Certain of the findings pointed toward the recognition of sea-floor spreading. During this period, Revelle himself was directly responsible for two major SIO studies.

One, on which he collaborated with Sir Edward "Teddy" Bullard and Arthur Maxwell, dealt with heat flow throughout the ocean floor. To their surprise, they found that it was the same at sea as on land. This unexpected result was eventually explained by the theory of plate tectonics and the research is widely credited with having helped lead to the theory.

The other study, on which he worked with Hans Suess, has proven to have particularly far-reaching consequences. In 1897, the Swedish chemist, Svante A. Arrhenius, had made a pioneering study of the relationship between global temperatures and atmospheric carbon dioxide. In 1957, Revelle and Suess revived Arrhenius's inquiry and published a paper observing that the steady increase of carbon dioxide from the burning of fossil fuels could pose a potentially serious danger. The oceans, they calculated, could not absorb more than half of the CO_2 resulting from the use of fossil fuels, and as more such emissions were produced, accumulations of CO_2 in the atmosphere could rise significantly in the coming decades. Warning that humanity was performing a "great geophysical experiment"—with the entire planet as its laboratory—they also contended that the phenomenon should be measurable with available means.

Revelle and Suess's 1957 paper served as a clarion call to the international scientific community to investigate the question of global warming, which also came to be referred to as "the greenhouse effect." In response to Revelle's lobbying, one of the projects undertaken during the 1957 International Geophysical Year (for which his main contribution was to plan the International Indian Ocean Expedition) was the setting up of CO_2 measuring stations at Mauna Loa. In 1990, *Scientific American* noted that "those measurements are one of the best-known curves in science. Each year has added another wiggle at the end of an increasingly upward slope—and the atmospheric CO_2 concentration has reached a level more than 350 parts per million, an 11 percent increase from 1957." As the curve has risen, so has public concern over global warming and over what can be done to monitor and control it. In 1977 Revelle chaired the Energy and Climate panel of the National Academy of Sciences that recommended an interdisciplinary investigation of CO_2 accumulation. One of his final papers, written in 1991, reviews suggestions for reducing and increasing absorption of CO_2 emissions.

During his tenure as director of SIO, Revelle's restless imagination was seized by another grand vision—a great research university that would affiliate with SIO and also with the University of California. No oceanographic program, he said, could be sure of maintaining intellectual excellence for more than a generation without an attachment to a great university, a judgment shared at SIO's Atlantic counterpart, the Woods Hole Oceanographic Institution, which subsequently formed an affiliation with MIT.

In typically entrepreneurial style, he announced the plan in 1956 and set about building support for it, first among local community leaders. With them he argued that if San Diego wanted to become more than a set of distant suburbs in some amorphous hinterland of Los Angeles, the city had to have a major research university which could attract high-technology industry— as Stanford and Berkeley had done for the Bay Area and Silicon Valley or as MIT and Harvard had within Boston's Route 128.

The site he thought the obvious candidate was readily available: some 1100 acres of largely undeveloped pueblo land owned by the city of San

Diego just to the north of SIO, parts of which had been used by the Army and Marine Corps for training purposes. The message was well-timed to appeal to San Diego's business leaders, inasmuch as the end of World War II had diminished the city's wartime bustle and left it with an economically imbalanced emphasis on defense industries vulnerable to cutbacks and redirections of military spending.

Fortuitously, Revelle's initiative coincided with the publication of a new master plan for the University of California which foresaw the need to establish two new major campuses in southern California. Revelle had in mind a major research university with a heavy concentration of graduate students. The proposal ran into opposition, however, from faculty at the two major research campuses of the University, UCLA and Berkeley.

Certain faculty reviewers raised objections to the inclusion of graduate studies. Revelle later roasted them—all but impugning their immediate ancestry—for trying to forestall competition:

> The committees on educational policy consisted almost entirely of professors from Berkeley and UCLA. They were experts at seeing clouds no bigger than a man's hand. It was clear to them that a new graduate school would draw money away from their own campuses; it might even attract outstanding scientists who would better serve mankind in Berkeley or Los Angeles. They thought it would be nice to have an undergraduate school at La Jolla, managed by a farm team of dedicated teachers, which could provide well-trained new graduate students for their own laboratories.

These aspersions were provoked by one review in particular. A UCLA faculty report recommended that the projected new campus, then known as UC La Jolla, be permitted to offer only lower division undergraduate courses at first, and only after a later review, to add upper division courses, but not a graduate program. SIO's faculty replied caustically that Scripps had been granting doctoral degrees when UCLA was still a teacher's college.

Fortunately, as Revelle later pointed out, a special subcommittee appointed by President Robert Gordon Sproul and including a number of eminent Berkeley faculty, saved the day, as well as honor of the UC faculty, when it examined the proposal for a campus that would be linked to SIO, and include a graduate program, and welcomed it with enthusiasm.

Revelle also acquired an important ally in Sproul's successor, Clark Kerr, but both of them ran into difficulty obtaining regental approval. The effort proved bruising—"more so," Kerr later recalled, "to Roger than to me."

One Regent in particular, Los Angeles oil magnate Edwin Pauley, raised an objection to the La Jolla location, proposing instead that the campus be located in Balboa Park. Revelle saw this as a thinly disguised maneuver designed to prevent the establishment of a new UC campus in San Diego. Pauley must have realized, Revelle thought, that in view of San Diego's continuing growth, its citizens would reject a proposal to take precious parkland for any other purpose. Even when the city was much smaller in population, in the 1920s, San Diegans had voted to give part of the park to the Navy for a military hospital but had afterward rejected a proposal to locate San Diego State College (later SDSU) in another part of the park. In the unlikely event that this site were approved by the voters, moreover, Revelle contended that it would cost the new university the critical advantage of proximity to SIO in attracting top-flight faculty.

Pauley criticized the proposal for the La Jolla site by pointing out that it lay under the flight path of jets operating out of the nearby Miramar Naval Air Station. To impress the Regents with the seriousness of the problem, he invited them, a few at a time, to his vacation home on the eighteen-acre Coconut Island near Oahu which he owned and which was not far from a landing strip used to train aircraft carrier pilots. While the mostly elderly Regents were visiting, Pauley arranged to have a flight of jet aircraft swoop low overhead at dusk, afterburners glowing. They were duly impressed, and according to a later account one of them even hit the beach. Revelle countered by producing studies showing that noise levels were even worse at other existing universities, but Pauley persisted with his objection.

The issue came to a dramatic climax at a meeting held at UC Davis in October of 1959. There an architect retained by Pauley asserted that the acoustic problem would make the cost of building a campus on the La Jolla site twenty percent more expensive that it would be in Balboa Park. In preparing Kerr for the meeting, however, Revelle had managed to get hold of a different report by the same architect in which he claimed that the acoustic problem of a nearby location for Scripps Hospital (under the same flight path) would not appreciably increase costs. When Kerr produced the letter, a chagrined Pauley asked the architect if he was indeed its author, and the even more embarrassed architect admitted he was, gravely damaging Pauley's case. Revelle drove home his advantage by reporting that the commanding officer at Miramar had agreed to cooperate by altering the flight patterns a few degrees to the north (over Del Mar, whose residents have no vote in San Diego elections). Pauley was neither amused nor appeased, but the other Regents were persuaded and the La Jolla site was approved. In a referendum, San Diego voters supported the transfer of the land to the university, and after some congressional resistance to the ceding of federal title was overcome, construction began on the La Jolla site.

After winning authorization, Revelle worked tirelessly to recruit eminent faculty, on the theory that a great research university had to be built from the top down—or, as he put it, by laying the roof first. Among those he persuaded to join the faculty were such distinguished scientists as Harold Urey, a Nobelist at the University of Chicago, Joseph Mayer and Maria Goeppert Mayer (who has to become a Nobel Laureate), James Arnold, David Bonner, Milton Bramlette, Keith Brueckner, Walter Elsasser, Martin Kamen, Walter Kohn, Bernd Matthias and Bruno Zimm. With their help, he set about recruiting others and planning the educational design of the campus, emphasizing the college system—a commitment he later admitted had not worked out quite as planned because the departments proved unwilling to surrender their power over curricula and appointment to the colleges. Although the plan was constantly revised, Revelle clung to the basic objective throughout: "to build a foundation which would become one of the great universities of the world."

Such a university, Revelle thought, must be complex and changing and the work of many hands and many generations. In a beautiful and apt simile, he compared the task to the building of a medieval cathedral:

> The university was the modern counterpart of the medieval cathedral, rising in the heart of the city and lifting the spirits of man, serving their aspirations and bringing reality to their ideals. The university must be diverse; like a cathedral, it should have many chapels. Just as the building of the cathedral absorbed the devotion and skill of unknown craftsmen, so the building of the university was a cause in which modern men could lose and find themselves. In education, the university, like the cathedral, should serve in appropriate ways human beings of all ages and conditions, not simply a privileged generation.

Revelle's belief that the university should be open to all was passionately held. In 1950, when fear of communist subversion was reaching near-hysterical proportions, especially in southern California, he supported faculty who refused to accept the Regents' insistence on a loyalty oath, and helped fashion an acceptable compromise. Later, when he discovered in recruiting faculty that despite Supreme Court rulings outlawing restrictive covenants, La Jollans continued to observe a "gentleman's agreement" to discriminate on grounds of religion and race in selling and renting property, he persuaded real estate agents and other leading members of the community to break with the custom. Long before it became fashionable, he was a champion of equal educational opportunity for women; and women who met him often remarked that he was one of the rare men of his generation who actually paid attention to what they said and not just how they looked.

The implacable enmity of Regent Pauley, who was a longtime chair of the Board of Regents, and very likely of another Regent rankled by Revelle's opposition to the loyalty oath, led to bitter disappointment, when Revelle was twice passed over for appointment as chancellor of UCSD. He was not actively considered either when the campus was opened or later when the first chancellor, Herbert F. York, was compelled to resign for reasons of health. Revelle had to be content with the special tribute of having UCSD's first college named in his honor. But he chafed at the circumstances in which he found himself, saying he would have preferred the responsibility to the honor. In 1961 he therefore took another leave from SIO to accept a two-year appointment as science adviser to Stewart Udall, Secretary of the Interior in the Kennedy-Johnson administration, afterward returning briefly to UCSD in 1963 to serve again as Director of SIO and to accept appointment by President Kerr to a specialty created post with no defined responsibilities as Dean of Research for the UC system.

While in Washington, Revelle had made more of his appointment to assist the Secretary of the Interior than the job appeared to offer. He used the opportunity as a springboard to a new career as a kind of scientific missionary to the developing nations. Asked by the President's Science Adviser, Jerome Wiesner, to head a team of academics to advise the government of Pakistan about problems of soil salinity, he directed a project which proved to be one of the great successes of the "Green Revolution," greatly improving agricultural productivity in the Indus Valley of Pakistan. For his leadership of the project, Revelle was awarded the Medal of Sitara-i-Imtiaz by the government of Pakistan. The Pakistan task force had included some Harvard faculty, who had been impressed by Revelle's leadership. As a result, two years later, in 1964, Revelle was invited to launch a new Center for Population Studies at Harvard, where he accepted appointment as Richard Saltonstall Professor of Population Policy, a chair he held for just over a decade. There he taught courses and organized research efforts on population and development. Among the undergraduates who studied under him was Albert Gore, who, as U.S. Senator from Tennessee, has made global warming a major political issue, crediting the course he took with Revelle with having alerted him to it.

Reflecting on his years at Harvard, Revelle pointed out that he had sought to give the Center a broader emphasis than population alone: "I was never a demographer or a family planning expert. I was concerned with resources— the kind of things we had done in Pakistan: a systematic approach to land, water and energy." Uncontrolled population growth, he thought, was more the consequence than the cause of underdevelopment: "When you educate and give people mobility and hope, you solve the problem—they'll have fewer children." The Center's building has since been named in his honor.

Upon retiring from Harvard in 1975, Revelle returned to UCSD to become Professor of Science and Public Policy and an affiliate of the newly established Department of Political Science. For the next fifteen years, he taught courses in marine policy and population and development, and continued to be active as a researcher, writer and adviser to governments, foundations, and colleagues. In the 1980s he was among the earliest to urge Chancellor Richard C. Atkinson and President David P. Gardner that the time was ripe to establish a new center for Pacific studies at UCSD—a proposal which eventually bore fruit in the creation of the Graduate School for International Relations and Pacific Studies.

In recognition of his scientific work, Revelle received many honors and awards. These include thirteen honorary degrees, the Albatross Medal of the Swedish Royal Society of Science and Letters, the Compass Award of the Marine Technology Society, and the Outstanding Achievement Award of the Climate Institute. His highest honors were the Vannevar Bush Medal and the Agassiz Medal of the National Academy of Sciences, the Bowie Medal of the American Geophysical Union, the Tyler Ecology Prize, the Balzan Foundation Prize for Environmental Science, and finally, in 1991, the National Medal of Science, awarded by President George Bush in a ceremony in the White House; the citation praised Revelle for "his pioneering work in the areas of carbon dioxide and climate modification, oceanographic exploration presaging plate tectonics, and the biological effects of radiation in the marine environment, and studies of population growth and global food supplies."

He served with distinction in many scientific organization and on countless committees. Elected a fellow of the National Academy of Sciences, he served as chairman of its Ocean Science Board. He was chosen president of the American Association for the Advancement of Science and vice-president of the American Academy of Arts and Sciences. He was elected to membership in the American Philosophical Society. First chairman of the Special (later Scientific) Committee on Oceanic Research, organized by the International Council of the Scientific Unions, he was also president of the first International Oceanographic Congress at the UN in 1959, and chairman of the U.S. delegation to the Intergovernmental Oceanographic Commission in 1962 which he had helped found in 1961. He was a longtime member of the prestigious Cosmos Club in Washington. Active in the Council on Foreign Relations, he attended many international meetings, including those of the Pugwash Conference on science and world affairs from 1958 to 1981 and others devoted to problems of developing countries. He served on advisory panels to the President's Science Advisory Committee, the House Committee of Science and Astronautics, the Naval Research Advisory Committee, the National Science Foundation, and the Agency for International Development.

Despite all his honors and accomplishments, Revelle himself claimed not to be well-educated or even overly bright, but although he meant to be sincerely self-critical rather than falsely modest, he did his own talents an injustice. Like Socrates' "profession of ignorance," these self-deprecating remarks were constantly belied by the way his mind would penetrate problems and assimilate information. He often showed flashes of insight and he pursued a style of work his colleague Teddy Bullard called "his Sherlock Holmes method" of considering every possible hypothesis until he could discard any that did not fit the data. The respect he commanded among his scientific peers enabled him to lead and represent them in efforts to promote conservation of the earth's resources and environment. In an age of specialists, he respected the need for focused inquiry but sought, like the "Interpreters of Light" in Francis Bacon's vision of the New Atlantis, to synthesize the findings of specialists and apply them for social benefit. In an obituary for the Independent of London, the oceanographer Henry Chamok spoke for many environmentalists when he noted that "for an inspired view on earth science, and on its repercussions on the human predicament, he was in a class of his own."

To his family he was an endearing husband, father and grandfather. Dinner at the Revelle home, family members have noted, was like a seminar, for which the children, in-laws and grandchildren were expected to be on their mettle and hold their own, but which was intended not merely to test, but to inspire and educate.

To his community, he was not only the founder of a university in which people could take increasing pride, but also a Town Councillor, San Diego Rotary's Man of the Year in 1990, and, with his wife, the donor to the San Diego Historical Society of an easement providing that the ocean view must be maintained in perpetuity on a choice piece of property owned by the Revelle family adjoining the La Jolla Museum of Contemporary Art in the La Jolla cultural zone. It is too little to add that he was a patron of the arts. The Revelles were to the arts in San Diego something of what the Medicis had been to those of Florence. Accordingly, the La Jolla Chamber Music Society's 1991 Summerfest and the 1991–92 season of the San Diego Symphony were dedicated to his memory. He was similarly honored by the La Jolla Playhouse and other cultural and civic groups.

Roger Revelle died at the age of 82 on July 15, 1991, of complications of cardiac arrest, almost two years after successful heart surgery. At a memorial service in the La Jolla Presbyterian Church attended by hundreds who knew and admired him, he was eulogized by family, friends and the former pastor of the church, who recalled lively debates with his free-thinking friend and parishioner. At a moving farewell tribute at SIO, he was warmly praised for his contributions to oceanography by the directors of SIO who succeeded him, William A. Nierenberg and Edward A. Frieman. On that occasion, Frieman announced that a ship for multi-disciplinary research to be commissioned in 1994 will be named in his honor the R/V Roger Revelle. His ashes were taken out to sea by the Scripps research vessel New Horizon. A campus-wide remembrance was convened in October and addressed by Chancellor Atkinson, President Emeritus Kerr, Walter Munk of SIO, and Justin Lancaster, who spoke on behalf of Revelle's students.

A large man, standing 6'4", with a handsome face, an irresistible grin, and a deep and gentle voice, he made an unforgettable impression. He was also well known among undergraduates for the squeak of his size-fifteen shoes. At UCSD, where for generations to come, faculty and students will build and rebuild the cathedral of learning he founded, and as he put it, "lose and find themselves" in the process, his spirit will live on. But of Roger Revelle himself it can truly be said, as Thucydides wrote of Pericles, that no one who comes after him can hope to fill his shoes.

Sanford Lakoff (chair)
Walter Munk
S. Jonathon Singer

ADDENDUM: REVELLE AT HARVARD AND LATER

Since the foregoing faculty minutes understandably pay scant attention to Roger Revelle's achievements after he left the Scripps Institution, we append a brief account of his major activities after his move to Harvard. Revelle's style and interests had begun to shift even before he left Scripps. His experience with the waterlogging study in West Pakistan had awakened his interest in the problems of the developing nations. It also persuaded him that he could make a contribution to solving these problems.

When he moved to Harvard in September 1964 to organize and direct its newly established Center for Population Studies, Revelle brought a novel point of view to population problems. Whereas population studies had traditionally meant, essentially, demography, Revelle's work in West Pakistan and India made him sensitive to the importance of the impacts of human populations on the ecology and physical conditions of their environments, and thereby on the livelihood and quality of life they can attain. Furthermore, Revelle's training as a natural scientist had equipped him for investigating the interactions between people and their environments. Accordingly, Revelle oriented his Center's program and much of his own work around the relationships between people and their environments. In his own words:

> The Harvard Center emphasized a different view, attempting to gain
> understanding of the consequences of population change on human

lives and societies, and of the biological, cultural, and economic forces that influence human fertility.

The principal lines of his activities after leaving Scripps can be sketched briefly:

1. *Global Warming.* The possibility that continuing to emit carbon dioxide without restraint might drastically alter the global climate was one of Revelle's lifelong concerns. His most influential paper in this area was probably the pioneering one he wrote with Hans Suess in 1957, while he was still Director of Scripps. Though few relevant data were then available, the paper called attention to the possibility that the continued accumulation of carbon dioxide in the atmosphere could raise the earth's average temperature, as Svante Arrhenius and T. C. Chamberlin had suggested nearly a century before. In the interim, the dramatic increase in the world's use of fossil fuels had converted Arrhenius and Chamberlin's theoretical possibility into an ominous threat. Revelle continued to study this problem and report his findings in at least a dozen articles published throughout the remainder of his life.

Very likely, the most significant of these later papers was Revelle's evaluation of the state of knowledge a quarter of a century later, in "Carbon Dioxide and World Climate," published in the *Scientific American* in 1982. There he concluded that, despite all the research, the consequences of continued discharge of carbon dioxide would lead to both destructive and constructive consequences, and, although he clearly felt that the destructive ones would most likely predominate, Revelle did not feel sure enough about the probable net effect to announce his conclusion. Instead, he ended the paper by noting that: "It would be prudent to begin thinking about what the changes might be and how humankind might best avoid or ameliorate the unfavorable effects and gain the most benefit from the favorable ones."

2. *The "Population Problem."* The Harvard Center was established in response to widespread concerns about the rapid growth of population in the United States and the world, and particularly in the less developed countries. Over the years Revelle published about thirty papers dealing with population projections, the likely consequences of continued growth of the world population at a rate of about 2 percent per year, and recommendations (relating largely to family planning, particularly in developing countries) for abating the growth rate.

Among the most interesting of these papers were the estimates of the size of the population that the planet could support, which Revelle prepared for the Bucharest World Population Conference in 1974. He derived his estimates by detailed calculations of the flow of solar energy, the extent of arable land on the planet, the annual recharge of fresh water available for plant growth, human nutritional requirements and related data. These considerations led him to the conclusion that the world could support a stationary population of 50 billion people at a moderate standard of living. He was far from feeling that so large a population would be desirable, however, and he estimated that living conditions would be far more agreeable with half that many people or fewer. Revelle was well aware that these estimates, although based on careful evaluation of the flow and utilization of solar energy, omitted many important considerations. In particular, they ignored social and political factors, including the effects of inequalities of income both between and within countries. But the estimates do reveal the limits that would be attainable in a world without practical impediments to perfectly efficient agricultural production and food distribution everywhere.

Beginning in 1970, Revelle worked with Rose E. Frisch, a geneticist, investigating the effects of diet and exercise on women's fecundity. Both inadequate nutrition and vigorous exercise tend to reduce a woman's total body weight and the proportion of that weight consisting of fat. Analysis of a variety of data showed that low body weight and low proportions of fat, jointly or together, tend to interfere with the inception of menarche in adolescents and the continuation of menstruation in adults. The survival value of a mechanism that reduces the likelihood that a woman will conceive when she is either too small or too deficient in fat for successful gestation is evident. Other investigators have elucidated the physiological processes by which weight and the fat composition of the body influence the menstrual cycle. Frisch recounts in detail these and related investigations in her paper in this volume.

3. *Developing Nations.* After his initial visit to Pakistan, Revelle devoted a great deal of his time and energy to the problems of developing nations and visited them, particularly the Indian Subcontinent, frequently. His dedication to the Subcontinent stemmed from his experience in leading the White House team to study the problems of agriculture in West Pakistan in 1960. This experience is discussed in Jerome Wiesner's memoir in this volume, and, from a different viewpoint, in my paper also here. A dozen or more of Revelle's papers are based on his experiences in the Subcontinent and other developing regions, including Egypt.

4. *Oceanography.* Revelle always considered himself an oceanographer, although he did not participate in any oceanographic expeditions after he left Scripps. He wrote, over the years, somewhat more than a dozen articles in the *Scientific American* and other influential journals on the importance of

oceanography and its contributions to understanding the geology of the planet and to solving the world's food problems. Revelle tirelessly advocated increased attention to oceanography, serving as chairman of the Ocean Science Board of the National Academy of Sciences, as president of the UN's first International Oceanographic Congress, as the first chairman of the Committee on Oceanic Research of the International Council of Scientific Unions, and in many other influential capacities.

5. *Teaching.* While at Harvard, Revelle pursued an active teaching schedule. He taught a large and popular undergraduate course in the causes and cures of environmental problems and was coeditor of a book of readings for the course, *The Survival Equation.* He was also a professor in the Department of Population Sciences in the Harvard School of Public Health, as well as chairman of the department.

PRINCIPAL PUBLICATIONS OF ROGER REVELLE

This bibliography is based on material provided by Ms. Deborah Day of the Scripps Institute for Oceanography. The listing provided by Ms. Day was updated and reorganized by Ms. Linda Klaamas for the Roger Revelle Memorial Symposium.

I. Books Edited

1. *America's Changing Environment.* (ed. with Hans Landsberg). Boston: Houghton Mifflin, 1970.
2. *Rapid Population Growth: Consequences and Policy Implications.* National Academy of Sciences Study Group, Roger Revelle, Chairman. Baltimore: The Johns Hopkins Press, 1971.
3. *Population and Social Change.* (ed. with David Glass). London: Edward Arnold, 1972.
4. *The Survival Equation: Man, Resources, and His Environment.* (ed. with Ashok Khosla, and Maris Vinovskis). Boston: Houghton Mifflin, 1972.

II. Papers and Articles

1. "Report on precipitate obtained by removal of carbon dioxide from sea water." Part Vl in "Calcium equilibrium in sea water," by Haldane Gee, et al. *Scripps Institution of Oceanography Bulletin,* vol. 3, no. 7, 1932, 188–190.

2. "Physico-chemical factors affecting the solubility of calcium carbonate in sea water." *Journal of Sedimentary Petrology,* vol. 3. no. 3, December 1934, 103–110.

3. "The buffer mechanism of sea water." (with E. G. Moberg, D. M. Greenberg, and E. C. Allen). *Scripps Institution of Oceanography Bulletin,* vol. 3, no. 11, 1934, 231–278.

4. "Preliminary remarks on the deep-sea bottom samples collected in the Pacific on the last cruise of the CARNEGIE." *Journal of Sedimentary Petrology,* vol. 5, no. 1, April, 1935, 37–39.

5. "Oceanographic work of *U.S.S. BUSHNELL* in the North Pacific." (In cooperation with U.S. Hydrographic Office) International Association of Physical Oceanography, vol. 2, 1937, 105.

6. "The water masses of the North Pacific." (with R. H. Fleming and E. G. Moberg). International Association of Physical Oceanography. *Proces-verbaux,* vol. 2, 1937, 106–107.

7. "The origin and characteristics of the waters off the coast of southern California." (with R. H. Fleming and E. G. Moberg), *ibid.,* 108.

8. "A comparison of the oxygen content and other chemical properties of the intermediate and deep waters of the Atlantic and Pacific." (with E. G. Moberg and R. H. Fleming), *ibid.,* 146.

9. "The distribution of dissolved calcium in the North Pacific." (with E. G. Moberg), *ibid.,* 153.

10. "The organic nitrogen content of marine sediments off the west coast of North America." (with E. G. Moberg, R. H. Fleming and K. Heusner), *ibid.,* 145.

11. "The colloidal fractions of Pacific deep-sea clays." *ibid.,* 158.

12. "Recent offshore sediments from southern California." *Geological Society of America Bulletin,* v. 49, 1938, 1897–1898.

13. "Coring operations off the California coast." (with F. P. Shepard, R. S. Dietz, and K. O. Emery), *ibid.,* 1900–1901.

14. "Sediments of the Gulf of California." *ibid.,* vol. 50, 1939, 1929.

15. "Current measurements near the sea bottom." (with F. P. Shepard), *ibid.,* 1929–1930.

16. "Ocean bottom currents off the California coast." (with F. P. Shepard and R. S. Dietz). *Science,* v. 89, no. 2317, May 26, 1939, 488–489.

17. "Physical processes in the ocean." (with R. H. Fleming). *Recent Marine Sediments,* American Association of Petroleum Geologists, September 1939, 48–141.

18. "Sediments off the California Coast." (with F. P. Shepard), *ibid.,* 245–282.

19. "Current measurements near the sea bottom." International Association of Physical Oceanography. *Proces-verbaux,* v. 3, 1940, 114–115.

20. "On vertical circulation in the Central North Pacific." *ibid.,* 164–165.

21. "Criteria for the recognition of sea water in ground water." *American Geophysical Union, Transactions,* v. 22, 1941, 593–597.

22. "Recent marine ecological investigations of paleontological significance at the Scripps Institution of Oceanography." (with M. L. Natland and S. C. Rittenberg). National Research Council, Division of Geology and Geography, Report [no. I] of the Subcommittee on the Ecology of Marine Organisms, May 3, 1941, 35–42.

23. "Calculation of sound ray paths in seawater." (with R. H. Fleming). Report no. PN-92. San Diego: U.S. Navy Radio and Sound Laboratory, 1942, 39 mimeo.—declassified.

24. "Report on Investigation of Surface Ship Wakes." (with N. J. Holter). Report no. S-3. San Diego: U.S. Navy Radio and Sound Laboratory, 1942, 38 mimeo.—declassified.

25. "Radar Wave Propagation." (with L. J. Anderson, J. B. Smythe and F. R. Abbott). Report no. WP2, San Diego: U.S. .Navy Radio and Sound Laboratory, 1942, 50 mimeo.—declassified.

26. "Soundings in the Gulf of California and off the west coast of Lower California in 1939." Scripps Institution of Oceanography, La Jolla. *Records of Observations,* v. 1, no. 2, May, 1943, 81–83.

27. "Current measurements off the California coast in 1938 and 1939." (with F. P. Shepard), *ibid.,* 85–87.

28. *Marine Bottom Samples Collected in the Pacific Ocean by the CARNEGIE on Its Seventh Cruise.* Publication no. 556. Washington, DC: Carnegie Institution of Washington, 1944, 180.

29. "Apparatus for rapid conductometric titrations. Determination of sulfate." (with Lloyd J. Anderson). *Analytical Chemistry,* vol. 19, no. 4, April, 1947, 264–268.

30. "Bikini revisited—preliminary results of the scientific reserve during the summer of 1947." *Science,* vol. 106, Nov. 28, 1947, 512–513.

31. "Harald Ulrik Sverdrup—An Appreciation." (with W. H. Munk). *Journal of Marine Research,* vol. 7, no. 3, 1948, 127–131; bibliography of Harald Ulrik Sverdrup (1914–1947), 132–138.

32. "Diffusion in Bikini Lagoon." (with W. H. Munk and G. C. Ewing). *American Geophysical Union, Transactions*, vol. 30, 1949, 59–66.

33. "Education and training for oceanographers." (with V. O. Knudsen, A. C. Redfield, and R. R. Shrock). *Science*, v. 111, no. 2895, June 23, 1950, 700–703.

34. "Paleoecology and deep sea exploration." National Research Council, Division of Geology and Geography. Committee on a Treatise on Marine Ecology and Paleoecology. Report 1950–1951, 1951, 57–61.

35. "Barite concretions from the ocean floor." (with K. O. Emery). *Geological Society of America Bulletin,* vol. 62, 1951, 707–724.

36. "Heat flow through the floor of the eastern North Pacific Ocean." (with A. E. Maxwell). *Nature*, vol. 170, no. 4318, August 2, 1952, 199–200.

37. "On the geophysical interpretation of irregularities in the rotation of the earth." (with W. H. Munk). *Royal Astronomical Society, Monthly Notices* —Geophysical Supplement vol. 6, no. 6, September 1952, 331–347.

38. "Sea level and the rotation of the earth." (with W. H. Munk). *American Journal of Science*, vol. 250, 1952, 829–833.

39. "Free diving: a new exploratory tool." (with W. N. Bascom). *American Scientist*, v. 41, no. 4, 1953, 624–627.

40. "The earth beneath the sea—geophysical exploration under the ocean." In *Modern Physics for the Engineer,* by Lollis N. Ridenour. New York: McGraw-Hill, 1954, 306–329.

41. "Heat flow through the Pacific Ocean Basin." (with A. E. Maxwell). Association internationale de seismologie et de physique de l'interieur de la terre. Publications du Bureau central seismologique international. Serie A: Travaux scientifiques, 19-memoires presentes a l'Assemblee de Rome, 1954, 394–405.

42. "Fluctuating fishery stocks: what we know about this world-wide riddle; Part II—Other contributing factors: the green pastures of the sea." National Canners Association Convention, 47th, 1954. Fisher Products Conference. Proceedings. In *National Canners Association, Information Letter,* no. 1472 (Convention issue), January 30, 1954, 11.

43. "A deep sounding from the southern hemisphere." (with R. L. Fisher). *Nature*, v. 174, no. 4427, September 4, 1954, 469–470.

44. "Pelagic sediments of the Pacific." (with M. Bramlette, G. Arrhenius, and E. D. Goldberg). In *Crust of the Earth,* Arie Poldervaart, ed. Geological Society of America Special Paper, no. 62, 1955, 221–235.

45. "Evidence from the rotation of the earth." (with W. H. Munk). *Annales de Geophysique*, v. 11, no. 1, Jan–Mar 1955, 104–108.

46. "The seasonal oscillation in sea level." (with June Patullo, Walter Munk and Elizabeth Strong). *Journal of Marine Research*, vol. 14, no. 1, June 30, 1955, 88–155.

47. "The trenches of the Pacific." (with R. L. Fisher). *Scientific American*, v. 193, no. 5, November ,1955, 36–41.

48. "On the history of the oceans." *Journal of Marine Research*, v. 14, no. 4, December 31, 1955, 446–461.

49. "Oceanography in the Pacific during the international geophysical year." Hawaiian Academy of Science. 31st Annual Meeting, 1955–1956. Proceedings, 1956, 14.

50. "Nuclear science and oceanography." (with T. R. Folsom, E. D. Goldberg, and John D. Isaacs). Geneva, International Conference on the Peacetime Uses of Atomic Energy, 1955. Proceedings, v. 13. New York, United Nations, 1956, 371–380.

51. "The ocean off the California coast." (with Paul Horrer). In *California and the Southwest,* Clifford Zierer, ed. New York: Wiley, 1956, 80–96.

52. "Heat flow through the deep sea floor." (with E. C. Bullard and A. E. Maxwell). In *Advances in Geophysics*, vol. 3. New York: Academic, 1956, 153–181.

53. "The biological effects of atomic radiation." Summary Report of the Committee on Effects of Atomic Radiation on Oceanography and Fisheries, 1956 (Roger Revelle, Chairman with others). Washington, DC: National Academy of Sciences—National Research Council 1956, 73–84.

54. "Turmoil in earth." *Science Newsletter,* vol. 69, no. 2, Jan. 14, 1956, 23.

55. "Sea, ice, and rainwater." *Saturday Review*, vol. 39, September 1, 1956, 41–44.

56. "The oceans and the earth." *Ciencia* v. 16, nos. 11/12, Nov/Dec., 1956, 290–291.

57. "Deep sea research as a cooperative enterprise." In *Symposium on Aspects of Deep-sea Research.* Washington, DC: National Academy of Sciences National Research Council, 1956. Proceedings, 1957, 163–171. (National Research Council Publication no. 471.)

58. "Bikini and nearby atolls, Marshall Islands. Chemical erosion of beach rock and exposed reef rock." (with K. O. Emery). U.S. Geological Survey. Professional Paper 260–T, 1957, 699–709.

59. "General considerations concerning the ocean as receptacle for artificially radioactive materials." (with M. B. Schaefer). In *The Effects of Atomic Radiation on Oceanography and Fisheries.* Washington, DC:

National Academy of Sciences National Research Council, 1957, 1–25. (National Research Council Publication no. 551.)

60. "Carbonates and carbon dioxide." (with R. Fairbridge). In *Treatise on Marine Ecology and Paleoecology*, vol. 1, 1957, Joel Hedgpeth, ed. Geological Society of America. Memoir 67, 239–296.

61. "Carbon dioxide exchange between atmosphere and ocean and the question of an increase of atmospheric CO_2 during the past decades." (with H. E. Suess). *Tellus,* vol. 9, no. 1, 1957, 18–27.

62. "Scientist and Politician." *Science,* vol. 125, no. 3257, May 31, 1957, 1078–1079.

63. "Man between two oceans." *Think*, v. 23, no. 5, July 1957, 18–23.

64. "Sun, sea, and air." *Oceanus,* v. 5, nos. 3/4, Summer/Autumn 1957, 26–36; also in *Smithsonian Institution Annual Report,* 1958, 251–260 (with discussion).

65. "Nature's great sea-air engine." *Science Digest* v. 42, October, 1957, 18–22.

66. "International cooperation in marine sciences." *Science,* v. 126, no. 3287, December 27, 1957, 1319–1323.

67. "The multiple functions of a graduate school." Conference, the Association of Princeton Graduate Alumni. 7th, Princeton University, 1958, 89–103.

68. "Oceanic research needed for safe disposal of radioactive wastes at sea." (with M. B. Schaefer). International Conference on the Peaceful Uses of Atomic Energy. 2nd, Geneva, 1958. Proceedings, v. 18, 1958, 362–370.

69. "Dynamics of the CO_2 cycle." In *Conference on Recent Research in Climatology.* Scripps Institution of Oceanography, La Jolla California 1957. Proceedings. Berkeley: University of California Statewide Committee on Water Resources, 1958, 93–105.

70. "Some aspects of deep-sea exploration." In *Symposium on the Physical and Earth Sciences,* honoring the twenty-fifth presidential year of Robert Gordon Sproul. Berkeley: University of California, 1958, 53–65.

71. "Marine resources." (with M. B. Schaefer). In *Natural Resources,* Martin Huberty and Warren L. Flock, eds. New York: McGraw-Hill, 1958, 73–109. (Also translated into Russian under auspices of Professor Lev Zenkevich in Priroda).

72. "Dubious harvest." (excerpt from "Effects of atomic radiation on oceanography and fisheries." with M. B. Schaefer). *Saturday Review,* v. 41, July 5, 1958, 43–44.

73. "How deep is the ocean and other riddles of the sea." *Navy,* vol. 1, August 1958, 11–19.

74. "Oceanography Program: first twelve months." (with Gordon G. Lill et al.). *IGY Bulletin,* no. 16, October, 1958. In *American Geophysical Union Transactions,* v. 39, no. 5, October, 1958, 1011–1017.

75. "International cooperation in the marine sciences." *ICSU Review,* vol. 1, no. 1, 1959, 27–39.

76. "Opening Address to the International Oceanographic Congress," August 30, 1959. *Oceanus,* v. 6, no. 3, March, 1960, 2–4.

77. "The biological effects of atomic radiation." Summary Report of the Committee on Oceanography and Fisheries, 1960 (Roger Revelle, Chairman, with others). Washington, DC: National Academy of Sciences—National Research Council, 1960, 69–90.

78. "Some recent lessons." *Bulletin of the Atomic Scientists,* v. 18, no. 1, January, 1962, 20–24.

79. "Sailing in new and old oceans." *American Institute for Biological Sciences Bulletin,* v. 12, no. 5, 1962, 45–47.

80. "Gases." (with H. E. Suess). In *The Composition of Sea Water,* M. N. Hill, ed. New York: *Interscience,* 1962–63, vol. 1, 1962, 313–321.

81. "International cooperation and the two faces of science." In *The American Assembly. Cultural Affairs and Foreign Relations,* Robert Blum, ed. Englewood Cliffs, New Jersey: Prentice Hall, 1963, 112–138.

82. *Report of the Committee on Natural Resources* (Roger Revelle, Chairman, with others). Federal Council for Science and Technology. Washington, DC: Govt. Printing Office, 1963.

83. "Natural resources policies and planning for developing countries" (with Joseph L. Fisher). In *Science, Technology, and Development: Volume 1, Natural Resources.* United States Papers prepared for the United Nations Conference on the Applications of Science and Technology for the Benefit of the Less Developed Areas. Washington, DC: Govt. Printing Office. 1963, 1–13. Abstracted in *Ekstics.* v. 16, no. 96, November, 1963; reprinted in *The Population Crisis and the Use of World Resouces,* Stuart Mudd, ed. for the World Academy of Art and Science, Bloomington: Indiana University Press, 1964, 407–422.

84. "Mission to the Indus." *New Scientist,* v. 17, no. 326, February 14, 1963, 340–342.

85. *Federal Water Resources Research Activities.* Report of the Task group on Coordinated Water Resources Research (Roger Revelle, Chairman, with others). Committee on Interior and Insular Affairs, U.S. Senate, 88th Congress, March, 1963, 213.

86. "Water." *Scientific American,* vol. 209, no. 3, September, 1963, 93–106. Reprinted in *Ekistics,* vol. 17, no. 98, January, 1964, 19–22; and in *Technology and Economic Development, a Scientific American Book,* New York: Alfred A. Knopp, 1963, 53–69.

87. "Water-resources research in the Federal Government." *Science,* v. 142, no. 3595, November 22, 1963, 1027–1033.

88. "Ocean resources, education, and research: The interaction of science and government." In *California and the World Ocean: Proceedings of the Governor's Conference.* Sacramento: Office of State Printing, 1964, 10–18.

89. "The role and effect of technology in the nation's economy—A review of the effect of government research and development on economic growth—Statement of Roger Revelle." Hearings before a Subcommittee of the Select Committee on Small Business. U.S. Senate, 88th Congress, First Session. Part 3, June 20, December 17 & 18, 1963. Washington, DC: U.S. Govt. Printing Office, 1964, 226–235.

90. *"Economic Benefits From Oceanic Research."* National Academy of Sciences Committee on Oceanography, Special Report, 1964. Introduction and Summary, Roger Revelle, 1–15; "Benefits and Costs," 7–12, "International Cooperation," 39–42, and Appendix, 47–50, Roger Revelle and M. B. Schaefer "Transportation." 28–36, Roger Revelle, Sumner Pike, and Harris Stewart, "Weather Forecasting." 33–39, Roger Revelle, and Columbus Iselin.

91. "Oceans, science, and men." *Impact of Science on Society,* vol. XIV, no. 3, UNESCO, 1964, 145–178.

92. "International cooperation in oceanography." In *Research in Geophysics. Volume 2, Solid Earth and Interface Phenomena,* Hugh Odishaw, ed. The National Academy of Sciences and the Massachusetts Institute of Technology, 1964, 565–576.

93. *Report on Land and Water Development in the Indus Plain* (Roger Revelle, Chairman, with others). The White House—Department of the Interior Panel on Waterlogging and Salinity in West Pakistan. Washington, DC: The White House, January, 1964, 151.

94. "Science and Technology in international development." *African Studies Bulletin,* January, 1964.

95. "A long view from the beach." *New Scientist,* v. 21, February 1964, 485–487. Reprinted in *The World in 1984,* v. 1, Nigel Calder, ed. Middlesex, England/Baltimore, Maryland: Penguin Books, 1965.

96. *An Assessment of Large Nuclear Powered Seawater Distillation Plants.* A Report of an Inter-agency Task Group. Washington, DC: Office of Science and Technology, March, 1964, 31.

97. "Environment land, air, water." New *Republic*, v. 151, November 7, 1964, 25–32.

98. "Science as an art." *Whitman College Bulletin*, no. 68, November, 1964, 18–22.

99. "A UNESCO-sponsored institute for natural resource analysis." Proceedings, 19th Pugwash Conference on Science and World Affairs. London: Pugwash Continuing Committee, 1965, 190–198.

100. *National Seminar on U.S. Food Policy in Relation to World Food Needs.* A Report of Proceedings (with R. W. Reuter). Seminar sponsored by Center for Research and Education, Estes Park, CO., with support of the Rockefeller Foundation, 1965, 59.

101. "Reconstruction of higher education in India." (for the Indian Education Commission), 1965, 18.

102. "California and the use of the ocean: a planning study of marine resources." (Roger Revelle, Chairman, with others). Prepared for California State Office of Planning by Institute of Marine Resources, University of California, 1965.

103. "Atmospheric carbon dioxide: carbon dioxide from fossil fuels—the invisible pollutant." (with W. Broecker, H. Craig, C. D. Keeling and J. Smagorinsky). In *Restoring the Quality of Our Environment.* Report of the Environmental Pollution Panel, President's Science Advisory Committee. Washington, DC: The White House, 1965, 111–133.

104. "The earth sciences and the federal government." In *Basic Research and National Goals.* A Report to the Committee on Science and Astronautics of the U.S. House of Representatives by the National Academy of Sciences, 1965, 237–255.

105. "Waterlogging and salinity in the Indus Plain: some basic considerations." (with R. Dorfman and H. Thomas). *Pakistan Development Review,* 1965, 331–370.

106. "Oceanography." In *Listen* to *Leaders in Science,* Childers, J. Saxon and A. Love, eds. New York: David McKay Company, 1965, 233–248.

107. "Water and agriculture." *Journal of Parliamentary Information* (India), 1965.

108. "UNESCO's science programme and present-day world problems." *UNESCO Chronicle,* vol. XI, no. 1, January, 1965, 33–37.

109. "The problem of people." *Harvard Today,* Autumn 1965, 2–9. Reprinted in *Nature and Science,* vol. 3, no. 11, 1966, 1–4; also in *Population Review,* Madras, India, vol. 10, no. 1, 1966, 17–23; and in *Family Planning in an Exploding Population,* by John A. O'Brien, New York: Hawthorne Books, Inc., 1968, pp. 15–25.

110. "Population and food supplies: The edge of the knife." Proceedings of the National Academy of Sciences, v. 56, no. 2, 1966, 328–351. Reprinted in *Prospects of the World Food Supply,* a Symposium under the Chairmanship of J. George Harrar, National Academy of Sciences, 1966, 24–47

111. "Overcoming the world food crisis." *Proceedings, 15th Pugwash Conference on Science and World Affairs.* London: Pugwash Continuing Committee, 1966, 262–282.

112. "Salt, water, and civilization." In *Food and Civilization: A Symposium,* S. M. Farber, N. L. Wilson and R. H. Wilson, eds. Springfield, IL: Charles C. Thomas, 1966, 83–104.

113. "Okeanographia—nauka planetarnaza (Oceanography—a planetary science)." *Nauka i Zhizn (U.S.S.R., Journal of Science and Life),* vol. 5, 1966, 46–50.

114. "The role of the oceans." *Saturday Review, XLIX,* 1966, 39–41.

115. "On the efficient use of High Aswan Dam for hydropower and irrigation." (with H. A. Thomas, Jr.). *Management Science,* 128, 1966, B296–B311.

116. "Can man domesticate himself?" *Bulletin of Atomic Scientists.* v. 22. 1966, 2–7. Reprinted in *American Journal,* VI:2, 1966, 151–162; also in *Nichibei Forum,* Tokyo, Japan, vol. 13, no. 1, 1967, 18–28.

117. "Pollution of the environment." *Bulletin of American Academy of Arts and Sciences,* vol. XIX, no. 6, 1966, 2–7. Reprinted in *American Scientist,* v. 54, 1966, 167A–170A.

118. "Just how 'Limitless' are the ocean's food supplies?" *Conservation Catalyst,* v. 1, no. 2, 1966, 2–5.

119. "Education and the world's population problem." American Association of School Administrators: Official Report, 1966, 124–140.

120. "Science and social change." In *Government, Science, and Public Policy,* a compilation of papers prepared for the 7th Meeting of Panel on Science and Technology of Committee on Science and Astronautics, U.S. House of Representatives, 89th Congress, 2nd Session, January, 1966, 31–39.

121. "World war on hunger." Statement to Committee on Agriculture. Hearings before Committee on Agriculture, U.S. House of Representatives, 89th Congress, 2nd Session, February 14–18, 1966. Part 1, 1966, 21–41.

122. "The changing races of the United States." *Journal of the American Medical Association,* August 22, 1966.

123. "Some scientific problems of international development." *Washington Colloquium on Science and Society,* Second Series, Morton Leeds, ed. Baltimore: Mono Book Corp., 1967, 22–32.

124. "The conquest of the oceans." In *The Control of Environment—A Discussion at the Nobel Conference,* John D. Roslansk, ed. Amsterdam: North Holland Publishing Co., 1967, 17–37.

125. "Pollution and cities." In *The Metropolitan Enigma: Inquiries into the Nature and Dimensions of America's "Urban Crisis."* James Q. Wilson, ed. Washington, DC: Chamber of Commerce of the United States, 1967, 78–121. Reprinted by the Harvard University Press, Fall, 1968, 91–134.

126. "International biological program." *Science,* vol. 155, no. 3765, 1967, 957.

127. "Fewer people? More food?" *International Science and Technology,* vol. 66, 1967, 70–79.

128. "Hungry passengers on the spaceship earth?" *Christian Comment,* vol. 81, 1967, 1–4.

129. "International cooperation in food and population." *International Organization,* v. XXII, no. 1, 1967, 362–391. Reprinted in *The Global Partnership,* Richard N. Gardner and Max F. Millikan, eds. New York: Frederick A. Praeger, 1968, 362–391.

130. "Feeding the world's hungry millions. Part II, an expert's answer." *Forbes Magazine,* v. 100, no. 5, 1967, 26.

131. "Population and Nutrional Demands." (with G. Goldsmith, C. L. Beale, J. W. Brackett, R. W. Engel, W. A. Gortner, O. C. Johnson, T. Myers, J. Milner and R. E. Shank). *The World Food Problem: A Report of the Panel on the World Food Supply of the President's Science Advisory Committee,* v. 2, no. 1, 1967, 1–135.

132. "Water and land." (with N. C. Brady, A. L. Brown, R. M. Hagen, A. C. Orvedal, D. F. Peterson, M. B. Russell, W. Thorne, and J. van Schilfgarde), *ibid.,* 405–469.

133. "Distribution of food supplies by level of income." (with Rose Frisch), *ibid.,* 43–54.

134. "Education for agriculture in India." *ibid.,* 217–234.

135. "Variations in body weights among different populations." (with Rose Frisch), *ibid.,* 141.

136. International Biological Program—"Statement of Roger Revelle." Hearings before the Subcommittee on Science, Research and Development of the Committee on Science and Astronautics, U.S. House of Representatives, 90th Congress, First Session, on H. Con. Res. 273, no. 6, May 9, June 6, July 12, August 3 and 9, 1967, 2–17.

137. "Population." *Science Journal,* October, 1967, 113–119.

138. "Outdoor recreation in a hyper-productive society." *DAEDALUS*, v. 96, no. 4, Fall, 1967, pp. 1172–1191. Reprinted in *America's Changing Environment*, Roger Revelle and Hans E. Landsberg, eds. Boston: Houghton Mifflin, 1970, 253–274.

139. "International cooperation and the two faces of science." In *Cultural Affairs and Foreign Relations*, Revised. Paul Braisted, ed. New York: The American Assembly, Columbia Books, 1968, 136–171.

140. "Oceanography." In *Toward the Year 2018.* New York: Foreign Policy Association, Cowles Publishing Co., 1968, 165–177.

141. "The quality of the human environment" In *Federal Programs for the Development of Human Resources,* Joint Economic Committee, 90th Congress. 2nd Session, 1968, 602–617.

142. "Prospects and possibilities in the human environment: a symposium." *Print*, 1968, 102.

143. "Too many born? Too many die, So Says Roger Revelle" (with Milton Viorst). *Horizon.* vol. X, no. 3, 1968, 32–37.

144. "World population growth and needed food supplies during the next twenty years." *Journal of the American Veterinary Medical Association,* v. 153, no. 12, 1968, 1840–1842.

145. "Can the poor countries benefit from the scientific revolution?" In *Applied Science and World Economy,* Ninth Meeting of the Panel on Science and Technology of Committee on Science and Astronautics, U.S. House of Representatives, 90th Congress, 2nd Session, January 23–25, 1968, 235–252.

146. "On technical assistance and bilateral aid." *Bulletin of the Atomic Scientists,* March, 1968, 17–19.

147. "Acceptance and response to award of Bowie Medal." *American Geophysical Union Transactions.* v. 49, no. 2, June, 1968, 433–473. "'Our Little Universities' Depend on Students, Faculty Working Together." *Triton Times,* June 14, 1968, 9.

148. "Human and natural resources—interaction and development in India and Pakistan." In *Science in India and Pakistan,* Ward Morehouse, ed. New York: Rockefeller University Press, Fall, 1968.

149. "Unity and fission in oceanography." *Presidential address to the General Assembly of the International Association of Physical Sciences of the Ocean.* Bern, Switzerland, November, 1967. Proces-verbaux, International Association of the Physical Sciences of the Ocean, vol. 8, Fall, 1968.

150. "The harvest of the sea and the world food problem." *Oceanus,* vol. XIV, no. 4, 1969, 1.

151. "The age of innocence and war in oceanography." *Oceans Magazine,* v. 1, no. 3, 1969, 5–16.

152. "The ocean." *Scientific American,* v. 221, no. 3, September, 1969, 55–65.

153. "The height and weight of adolescent boys and girls at the time of peak velocity of growth in height and weight: longitudinal data." (with Rose Frisch). *Human Biology,* December, 1969, vol. 41, 531–559.

154. "Variation in body weights and the age of the adolescent growth spurt among Latin American and Asian population in relation to calorie supplies." (with Rose E. Frisch). *Human Biology,* December, 1969, vol. 41, 185–212.

155. "Ethics, environment and population." *The New York Times,* January, 1970.

156. "Population and food in east Pakistan." (with H. A. Thomas, Jr.). *Industry and Tropical Health,* v. VII, 1970, 27–42.

157. "Height and weight at menarche and a hypothesis of critical body weights and adolescent events." (with Rose E. Frisch). *Science,* v. 169, 1970, 397–399.

158. "The height and weight of girls and boys at the time of the initiation of the adolescent growth spurt in height and weight and the relationship to menarche." (with Rose E. Frisch). *Human Biology,* vol. 43, no. 140, 1971.

159. "Height and weight at menarche and a hypothesis of menarche." (with Rose E. Frisch). *Archives of Disease in Childhood,* v. 46, no. 695, 1971.

160. "Folgen des raschen Bevoelkerungswachstums in den Entwicklungslaendern" ("Consequences of Rapid Population Growth in the Poor Countries"). *UMSCHAU in Wissenschaft und Technik* January 4, 1971.

161. "Paul Ehrlich: new high priest of ecocatastrophe." *Family Planning Perspectives,* April, 1971.

162. "The population dilemma: people and behavior." *Psychiatric Annals,* vol I, no. 1, September, 1971 (entire issue).

163. "Population growth and environmental control." (with H. A. Thomas, Jr.). Proceedings of ECAFE Conference on Population and Environment, Bangkok, September, 1971.

164. "Population projections for Bangladesh." (with H. A. Thomas, Jr., R. Tabors and F. Benford). Harvard Center for Population Studies, 1972, iii–110.

165. "Some consequences of rapid population growth." In *Are Our Descendants Doomed? Technological Change and Population Growth,* Harrison Brown and Edward Hutchings, Jr., eds. New York: Viking Press, 1972, 42–71.

166. "Reply to Holdren." *Family Planning Perspectives,* April, 1972.

167. "Possible futures for Bangladesh." *Asia,* Spring, 1973.

168. "Freedom of Oceanic Research." Editorial, *Science,* vol. 181, no. 4098, August 3, 1973, 393.

169. "Components of weight at menarche and the initiation of the adolescent growth spurt in girls: estimated total water, mean body weight and fat." (with Rose E. Frisch and Sole Cook). *Human Biology,* vol. 45, no. 1 September 1973, 469–483.

170. "Will the earth's land and water resources be sufficient for future populations?" Background Paper for United Nations Symposium on Population, Resources, and the Environment, Stockholm, October, 1973.

171. "The balance between aid for social and economic development and aid for population control." *Journal of International Health Services,* November, 1973.

172. "Rich nations and poor: insights into world population problems." *Interview,* November, 1973.

173. "Food and population." *Scientific American,* v. 231, no. 3, September, 1974, 160–170. Reprinted in *The Human Population,* San Francisco: W. H. Freeman & Co., 1974, 118–128.

174. "Will the earth's land and water resources be sufficient for future populations?" In *The Population Debate: Dimensions and Perspectives.* Papers of the World Population Conference, Volume II, Bucharest, 1974. New York United Nations, 1975, 3–14.

175. "Population and Environment." and "Population and Resources." background papers for United Nations Population Conference, Bucharest, August, 1974.

176. "The ghost at the feast." *Science,* v. 186, November 15, 1974, 589.

177. "Science and politics." Round-table discussion. *Harvard Political Review,* Winter, 1974.

178. "Population and food problems can be solved." The *Boston Globe,* December, 1974.

179. "The Ganges water machine." (with V. Lakshminarayana). *Science,* v. 188, 1975, 611–617.

180. "The scientist and the politician." *Science,* v. 187, March 21, 1975, 1100–1105.

181. "Emilio Q. Daddario, President Elect." *Science,* v. 192, 1976, 270–271.

182. "Die Zukunft unseres Planeten: grossartige neue Welt oder Alptraum?" ("Science and Technology for the Poor Countries"). *Umschau in Wissenschaft und Technik,* v. 76, no. 13, 1976, 422–425.

183. "Energy use in rural India." *Science,* v. 192, 1976, 967–975.

184. "The resources available for agriculture." *Scientific American,* September, 1976, 165–178.

185. "Flying beans, botanical whales, Jack's beanstalk and other marvels." In *Report of the National Research Council,* 1977, 173–200.

186. "Energy and rural development" In *The Many Facets of Human Settlement— Science and Society,* Irene Tinker and Mayra Bubinec, eds. New York: Pergamon Press, 1977, 133–142. Reprinted in *Habitat,* vol. 1, no. 2, Pergamon Press, 1977.

187. Seminar on Development of Small Scale Hydroelectric Power and Fertilizer Production in Nepal, February 28–March 3, 1977 (with Douglas Smith). The Asia Society, New York, 1977, iv & 15.

188. "Technology and the poor." In *Technology and Society,* Oakridge Bicentennial Lectures. Oakridge: Oakridge National Laboratory, 1977, 60–86.

189. "Changing numbers of mankind." (with W. W. Howells). In *Human Biology and Ecology,* by Albert Damon. New York: W. W. Norton and Company, 1977, 304–325.

190. "The carbon cycle and the biosphere." (with Walter Munk). *Energy and Climate,* National Academy of Sciences, November 1977, 140–158.

191. "Overview and recommendations." *ibid.,* 1–31.

192. "Columbus O'Donnell Iselin II (1904–1971)." *American Philosophical Society, Year Book 1977,* Philadelphia, PA, 1978, 61–71.

193. *The Agricultural Potential and Long Term Stability of the Indus Basin- Some Questions for Research* (with Ralph Cummings, Walter P. Falcon, and Rodney Tyers). A Report to the Planning Commission of Pakistan, June 1978. Islamabad. The Ford Foundation, 1978, 168.

194. "Requirement for energy in the rural areas of developing countries." In *Renewable Energy Resources and Rural Applications in the Developing World,* Norman Brown, ed. American Association for the Advancement of Science, 1978, 11–26; 150–160.

195. "Energy and climate." (with Donald C. Shapero). *Environmental Conservation,* v. 5, no. 1, 1978, 1–11 .

196. "The past and future of ocean drilling." Joint Oceanographic Institutions Incorporated, Washington DC, 1978. Reprinted in *Water Spectrum,* v. 12, no. 4, Winter 1979–80, 44–51; and in *The Deep Sea Drilling Project: A Decade of Progress,* R. G. Douglas and E. L. Winterer, eds. Tulsa, OK: Society of Economic Paleontologists and Mineralogists, October, 1981, 1–4. (Special Publication no. 32.)

197. "Energy sources for rural development." *Energy,* v. 4, 1979, 969–987.

198. "Margaret Mead—an American phenomenon (1901–1978)." *Science,* 1979.

199. "Postscript: Population growth and energy use. E. F. Schumacher as prophet." *Population and Development Review,* v. 5, no. 3, September 1979, 542–544.

200. "The [Woods Hole] Oceanographic [Institute] and how it grew." In *Oceanography. The Past,* M. Sears and D. Merriman, eds. Springer Verlag, NY: 1980, 10–24.

201. "Energy and development—the case of Asia." In *Science and Ethical Responsibility,* S. D. Lakoff, ed. Reading, MA: Addison Wesley Publishing, 1980, 287–303.

202. "Energy and food in the poor countries." In *Energy and the Developing Nations,* Peter Auer, ed. Proceedings of Electric Power Research Institute Workshop, New York, March, 1980. New York: Pergamon Press, 1981, 110–122.

203. "Climate and the ocean." *Edis,* vol. 11, no. 4, July 1980, 3–9.

204. "Energy dilemmas in Asia: the needs for research and development." *Science,* vol. 2099, July, 1980, 164–174.

205. "Statement of Dr. Roger Revelle on International Applications of Renewable Energy Resources." Subcommittee on Energy Conservation and Supply, Committee on Energy and Natural Resources of the U.S. Senate. 96th Congress, August 19 and September 5, 1980. Publication no. 96–147. Washington, DC: U.S. Gov't. Printing Office, 1980, 3–20.

206. "Biological research and the Third World countries." *Bioscience,* November, 1980.

207. Introduction to *The Oceanic Lithosphere,* v. 7 of *The Sea,* C. Emiliani, ed. New York: John Wiley & Sons, 1981, 1–17.

208. "Environmental and socio-economic research needed for better appraisal of the CO_2 problem." Statement of Dr. Roger Revelle in *Carbon Dioxde and Climate: The Greenhouse Effect.* Subcommittee on Natural Resources, Agricultural Research and Environment, and Subcommittee on Investigations and Oversight. Committee on Science and Technology, U.S. House of Representatives, 97th Congress, July 31, 1981 (no. 45). Washington, DC: U.S. Gov't. Printing Office, 1981, 7–26.

209. "Marine technical cooperation in the 1980s: An overview." In *International Cooperation in Marine Technology, Science and Fisheries: The Future U.S. Role in Development.* Proceedings of Workshop, January, 1981. Washington, DC: National Academy of Sciences, May, 1981, 11–30.

210. "International foundation for science." *Bulletin of the Atomic Scientists,* v. 37, no. 9, November, 1981, 27–32.

211. "Resources." In *Population Growth and World Economic Development,* J. Faaland, ed. Norwegian Nobel Foundation 1982, 50–77.

212. "Review of Building a Sustainable Society." by Lester R. Brown. *Population and Development Review,* vol. 8, no. 4, The Population Council, 1982, 829–834.

213. "Marine scientific and technical cooperation in the 1980s." *International Cooperation in Marine Science and Technology in the Pacific Region,* Ocean Assn. of Japan, Tokyo, March, 1982, 4–10

214. "The oceans and climate—the need for international cooperation." *International Cooperation in Marine Science and Technology in the Pacific Region,* Ocean Assn. of Japan, Tokyo, March, 1982, 48–53.

215. "Basic research for energy." In *Proceedings of Conference on Basic Energy Research,* National Science Foundation, March 13, 1982.

216. "Science, commerce, and resources in the ocean." *The Center Magazine,* v. XXV, no. 2, March–April, 1982, 53–54.

217. "El mar y el clima." (with Bertrand Thompson). *Ciencia y Desarolla,* v. XIII, no. 43, March–April, 1982. English version: "The oceans and climate." (with Bertrand Thompson). *Impact of Science on Society,* v. 32, no. 3, July–September, 1982, 271–280.

218. "Carbon dioxide and world climate." *Scientific American* vol. 247, no. 2, August, 1982, 35–43.

219. "Effects of a carbon dioxide-induced climatic change on water supplies in the Western United States." (with Paul E. Waggoner). In *Changing Climate.* National Academy Press, 1983, 419–432.

220. "Methane hydrates in Continental Slope sediments and increasing atmospheric carbon dioxide." In *Changing Climate.* National Academy Press, 1983, 252–261.

221. "Probable future changes in sea level resulting from increased atmospheric carbon dioxide." In *Changing Climate.* National Academy of Sciences, National Academy Press, 1983, 433–448.

222. "Annual report of informal meeting on CO_2 and the Arctic Ocean." In *Changing Climate.* National Academy of Sciences, National Academy Press, 1983, 483–486.

224. "World climate research program—oceanography." *UNESCO International Marine Science Newsletter,* no. 34, 1983, 3–4.

225. *Exchange of CO₂ Between the Atmosphere and the Biosphere: Stimulation of the Terrestrial Vegetation by Fossil Fuel CO₂ in Connection with Other Manmade Changes* (with G. Kohlmaier and C. D. Keeling). Report to U.S. Department of Energy, 1983.

226. "The ocean and the carbon dioxide problem." *Oceanus,* v. 26, no. 2, Summer, 1983, 3–9.

227. "Soil dynamics and sustainable carrying capacity of the earth." In *Global Change, a Symposium of the International Council of Scientific Unions,* T. F. Malone and J. G. Roederer, eds. ICSU Press Symposium Series no. 5, 1984, 341–349.

228. "The problem with carbon dioxide." In *1984 Yearbook of Science and the Future.* Chicago: Encyclopaedia Britannica, Inc., 1984, 131–143.

229. "The world supply of agricultural land." In *The Resourceful Earth,* J. L. Simon and H. Kahn, eds. Oxford: Basil Blackwell Inc. 1984, 184–201.

230. "Atmospheric CO_2 changes recorded in lake sediments." (with D. Lal). *Nature,* vol. 308, no. 5957, March 22, 1984, 344–346.

231. "Present and future state of living marine and fresh water resources." In *The Global Possible: Resources, Development, and the New Century,* Robert Repetto, ed. New Haven: Yale University Press, 1985, 431–456.

232. "Carbon dioxide and other greenhouse gases in the ocean, atmosphere and biosphere, and future climatic impacts." In *Chemical Events in the Atmosphere and Their Impact on the Environment,* G. B. Marini-Bettolo, ed. Citta del Vaticano: Pontificia Academia Scientiarum, 1985, 405–428.

233. "The need for international cooperation in marine science and technology." In *Ocean Yearbook,* E. Mann Borgese and N. Ginsburg, eds. Chicago: University of Chicago Press, 1985, 130–149.

234. "An unconventional approach to integrated ground and surface water development." (with V. Lakshminarayana). In *Water Resources Systems Planning,* M. C. Chaturvedi and P. Rogers, eds. Bangalore: Indian Academy of Sciences, 1985, 147–157. Reprinted in *Sadhana,* Academy Proceedings in Engineering Sciences. v. 8. Part 2, 1985, 147–157.

235. "Toxins in the human food system." In *Basic and Applied Mutagens,* A. Muhammed and R. C. von Borstel, eds. New York: Plenum Press, 1985, 11–26.

236. "Effects of el nino/southern oscillation on the atmospheric content of carbon dioxide" (with C. D. Keeling). *Meteoritics, The Journal of the Meteoritical Society,* v. 20, no. 2, Part 2, 1985, 437–450.

237. "Review of *The Bad Earth: Environmental Degradation in China,* by Vaclav Smil and *The State of India's Environment 1982: A Citizen's Report.* In *Population and Development Review,* v. 11, no. 2, 1985, 348–354.

238. "Oceanography in Space." *Science,* v. 228, April 12, 1985, 133.

239. "Hydrology and climate." In *U.S. Geological Survey, Open-File Report 86-540.* Proceedings. Fifth National Conference. Water Resources Division, November 17–21, 1985 (J. H. Green, J. E. Moore, A. W. Spieker, eds.), San Diego, California, 1986.

240. "Global ocean monitoring for the world climate research programme" (with F. Bretherton). *Environmental Monitoring and Assessment,* no. 7, 1986, 79–90.

241. "H. W. Menard. 1920–1986." *EOS,* v. 67, no. 18, May 6, 1986, 441–446.

242. "How I became an oceanographer and other sea stories" *Ann. Rev. Earth Planet Science,* v. 15, 1987, 1–23.

243. "Modelling stimulation of plants and ecosystem response to present levels of excess atmospheric CO_2" (with G. H. Kohlmaier, H. Broehl, E. O. Sire, and M. Ploechi). *Tellus,* 1987, 39B, 155–170.

244. *Perspectives on the Crisis of UNESCO.* Report of a conference, Rancho Santa Fe, California, January 31–February 2, 1986 (with W. Kohn and F. Newman, ed). University of California, Institute on Global Conflict and Cooperation, 1987.

245. "Food, population and conflict in Africa." In *Global Problems and Common Security: Annals of Pugwash 1988,* J. Rotblat and V. I. Goldanskii, eds. New York: Springer, 1989, 240–247.

246. "Energy options and climatic effects." (with D. Burns). In *Climate and Development: Climatic Change and Variablity and the resulting social, economic and technological implications,* H. J. Karpe, D. Otten, S. C. Trinidade, eds. New York: Springer in association with the United Nations, 1990, 261–69.

III. Articles in Press

1.   "How should oceanographers be educated?" *Guide to U.S. Academic Programs in Oceanography and Related Fields, 1978–1989.* Washington DC: Marine Technology Society.

2.   "Modelling the seasonal contribution of a CO_2 fertilization effect of the terrestrial vegetation to the amplitude increase in atmospheric CO_2 at Mauna Loa Observatory." (with G. H. Kohlmaier, E. O. Sire, A. Janacek, C. D. Keeling, and S. C. Piper). *Tellus.*

APPENDIX 4

ROGER REVELLE MEMORIAL SYMPOSIUM PARTICIPANT LIST

Bill Alonso
Jim Anderson
Carmen Barroso
David Bell
Gretchen and Warren Berggren
Ramesh Bhatia
Peter S. Bing
Lew Branscomb
John Briscoe
The Hon. D. Allan Bromley
Harvey Brooks
David Burns
Robert Burden
James Butler
Tom Cabot
Lincoln Chen
Hollis B. Chenery
Bill Clark
Kingsley Davis
Paul Demeny

John Dixon
Robert Dorfman
Robert Duemling
Arthur Dyke
Nicholas Eberstadt
Myron Fiering
Harvey Fineberg
Rose Frisch
Bob Frosch
Dick Gamble
Edward Goldberg
Joe Harrington
Allan Hill
Heinrich Holland
Mr. and Mrs. Gary C. Hufbauer
Howard Hiatt
Stein Jacobsen
Henry Jacoby
Milton Katz
David Keeling

Ashok Khosla
Nathan Keyfitz
Justin and Star Lancaster
Hans Landsberg
George Lewis
Harvey Liebenstein
Sandy Lieberman
The Hon. Robert S. MacNamara
Tom Malone
Steve Marglin
George Masnick
Jean Mayer
Jim McCarthy
Michael McElroy
Geoffrey McNicoll
Ralph Mitchell
Walter Munk
Christopher Murray
Mrs. Mary Paci
Roland F. Pease, Jr.
Ralph Potter
Don Price
Victor Rabinowitz
Howard Raiffa
Robert Repetto
Ellen Revelle
Mr. and Mrs. William Revelle
Alan Robinson
Peter Rogers

Hilton Salhanick
Ruth Scheer
Nevin Scrimshaw
Gita Sen
Mrs. Anne Shumway
Frederick Singer
Gene Skolnikoff
Douglas Smith
John C. Snyder
Rob Stavins
Peter Stone
Dick Tabors
Mrs. Ping Tai
Harold Thomas
Vernon Thomas
Mrs. Tyler
Maris Vinoskis
Paul Waggoner
Tom Weller
Alvin Weinberg
Jerome Wiesner
Dudley Willis
Richard Wilson
Wilma Winters
George Woodwell
John Wyon
Wassim Zaman
George Zeidenstein

REVELLE SYMPOSIUM AUTHORS' BIOGRAPHIES

PAUL DEMENY is Distinguished Scholar at the Population Council and has been Editor of the quarterly journal *Population and Development Review* since its inception in 1975. He is a Fellow of the American Association for the Advancement of Science and a former President of the Population Association of America. He graduated from the University of Budapest, and studied at the Institut universitaire de hautes etudes internationales in Geneva, Switzerland, and received a Ph.D. in economics from Princeton University. Dr. Demeny has served on the economics faculties of Princeton University, the University of Michigan and the University of Hawaii. He also taught at the University of California, Berkeley and was founding Director of the East-West Population Institute at the East-West Center in Honolulu, Hawaii. Dr. Demeny has also served as Vice President of the Population Council and as Director of the Demographic Division and the Center for Policy Studies at the Council. He has published extensively on the study of population dynamics and economic growth, with particular emphasis on the relationships between demographic change and economic and social development, and related issues of public policy.

ROBERT DORFMAN retired several years ago as David A. Wells Professor of Political Economy at Harvard University. Early in his long career, he pioneered in the introduction of linear programming methods in economics and wrote, in collaboration with Professors Samuelson and Solow of M.I.T., *Linear Programming and Economic Analysis,* which became the definitive treatise on that topic. Later, he became interested in natural resources and environmental economics, and collaborated with professors in several other Harvard departments in another influential interdisciplinary treatise, *Design of Water Resource Systems.* He has continued to be concerned with natural resource problems, and is currently engaged in an international study on the use of the water of the Jordan River by Israel, Jordan and the Palestinian People. Professor Dorfman has written many scholarly papers on the foregoing and other technical topics in economics, and has delivered lectures around the world. He has served as President of the Institute of Management Sciences and Vice President of the American Economic Association. In 1993 the American Economic Association elected him a Distinguished Fellow, a distinction awarded to at most two economists in a year.

NICHOLAS EBERSTADT is a visiting fellow at the Harvard University Center for Population and Development Studies. He has been a consultant to the World Bank, the State Department, the Agency for International Development and the Census Bureau. He has taught several courses at Harvard University on population and natural resources, agricultural economics, social science and social policy and problems of policy-making in developing countries. His books include *The Poverty of Communism* (Transaction Books 1988), *Foreign Aid and American Purpose* (AEI 1989), *The Population of North Korea* (University of California 1992), *Korea Approaches Reunification* (M. E. Sharpe 1995) and *The Tyranny of Numbers: Mismeasurement and Misrule* (AEI 1995). He has written several books and numerous articles on population and poverty. Mr. Eberstadt received an A.B. from Harvard College, a M. Sc. from the London School of Economics, and an M.P.A. from the Kennedy School of Government at Harvard University.

ROSE E. FRISCH received her B.A. from Smith College, a Masters in Zoology from Columbia University and a Ph.D. in Genetics from the University of Wisconsin. She is an Associate Professor Emerita for Population Sciences in the Department of Population Sciences and International Health at the Harvard School of Public Health, and has been a Member of the Harvard Center for Population and Development Studies since its inception. Professor Frisch is a reproductive biologist whose research over the last decades has focused on the biological and environmental determinants of female fertility. She was a Fellow of the John Simon Guggenheim Memorial Foundation, a Sigma Xi National Lecturer, and a Fellow at the Bunting Institute. She is the author of over 100 journal articles, and the book *Adipose Tissue and Reproduction* (Karger, Basel 1990).

EDWARD D. GOLDBERG retired from the Scripps Institution of Oceanography in 1994, where he held various positions including Professor of Chemistry. He has also held visiting faculty positions at the University of Washington, the University of Chicago, Harvard University and Otago University in Dunedin, New Zealand. Professor Goldberg received the B. H. Ketchum Award in 1984 from the Woods Hole Oceanographic Institution and in 1989 he was awarded the Tyler Prize for Environmental achievements. Professor Goldberg received a B.S. from the University of California, Berkeley and a Ph.D. in chemistry from the University of Chicago. Professor Goldberg's scientific interests include the geochemistry of natural waters and sediments, the demography of the coastal zone, waste management and marine pollution. Professor Goldberg is a member of the National Academy of Sciences and is a fellow of the American Association for the Advancement of Science and the American Geophysical Union. He has published extensively and served as editor of a technical series in oceanography, *The Sea: Ideas and Observations.* He was coeditor of the volumes *Earth Sciences and Meteorites* and *Man's Impact on Terrestrial and Marine Ecosystems.* His latest book is *Coastal Zone Space* (UNESCO Publishing 1994).

NATHAN KEYFITZ has been a Professor of Sociology at the University of Toronto, the Universite de Montreal, the University of Chicago, the University of California at Berkeley and Harvard University. He also served as the Director of the Harvard Center for Population and Development Studies. Professor Keyfitz worked for extended periods in Indonesia, Sri Lanka and India. He is a member of the U.S. National Academy of Sciences and the Royal Society of Canada, and has received honorary degrees from eight universities. He has published extensively on the mathematics of population and the interconnections of population, development and environment.

HANS H. LANDSBERG is Senior Fellow Emeritus and Consultant-in-Residence in the Energy and Natural Resources Division of Resources for the Future in Washington, DC, where he has also held the positions of Director of the Energy and Natural Resources Division and Director of the Resource Appraisal Program. Mr. Landsberg studied law at the Universities of Freiburg,

Heidelberg, Berlin, and received a B.Sc. in Economics from the London School of Economics and an M.A. in Economics from Columbia University. He is a Fellow of the American Academy of Arts and Sciences, and of the National Science Foundation, and was the 1984 recipient of the annual award of the International Association for Energy Economics. Mr. Landsberg has published extensively on several topic relating to resources and the environment, and his books include *Resources in America's Future* (Johns Hopkins 1963), *Energy Today and Tomorrow: Living with Uncertainty* (Prentice-Hall 1983), and *Competitiveness in Metals: The Impact of Public Policy* (Mining Journal Books 1992).

Massimo Livi-Bacci is a Professor of Demography in the Faculty of Political Sciences in the University of Florence. He is a past President of the International Union for the Scientific Study of Population and currently holds the title of Honorary President. He has lectured widely in Europe and the Americas and has written several books and articles on fertility, nutrition and population issues, including *A Century of Portuguese Fertility* (Princeton University Press 1971), *A History of Italian Fertility during the Last Two Centuries* (Princeton University Press 1977), *Population and Nutrition* (Cambridge University Press 1990) and *A Concise History of World Population* (Blackwell 1992). His research interests include historical as well as contemporary population issues and furthering and developing the study of links between demography and biological and social sciences.

Geoffrey McNicoll received a B.Sc. from the University of Melbourne, Australia and a Ph.D. from the University of California, Berkeley. He is a Professor at the Research School of Social Sciences at the Australian National University and Senior Associate at the Population Council. Professor McNicoll's research interests include the consequences of population changes, comparative analysis of demographic regimes, institutional change and population policy. He has written numerous books and articles including *Trade and Growth in the Philippines* (Cornell University Press 1971), *Fertility Decline in Indonesia: Analysis and Interpretation* (National Academy Press 1983) and *Rural Development and Population* (Oxford University Press 1990).

Walter Munk is a Professor of Geophysics at the Scripps Institution of Oceanography at the University of California in San Diego. He served as the first Director of the La Jolla Laboratory of the Institute of Geophysics and Planetary Physics at the Scripps Institution. He is a member of the National

Academy of Sciences, the Royal Society of London and the Russian Academy of Sciences. He has received numerous awards, including the National Medal of Science, the Vetlesen Prize from Columbia University, the Agassiz Medal from the National Academy of Sciences, the William Bowie Medal from the American Geophysical Union and the Presidential Award of the New York Academy of Sciences. His most recent book, *Ocean Acoustic Tomography* (Cambridge University Press 1995) was written with Worce... r and Wunsch.

PETER ROGERS is the Gordon McKay Professor of Environmental Engineering and Professor of City Planning at Harvard University. He was Visiting Professor at Technische Hochschule, Vienna, and at the Indian Institute for Technology, New Delhi. He is the recipient of Guggenheim and Twentieth Century Fellowships. He worked closely with Roger Revelle on population, water resources, and environmental problems in Pakistan, India and Bangladesh from 1963 until Revelle left Harvard in 1975. Professor Roger's research interests include the consequences of population on natural resources development, conflict resolution in international river basins, and the development of indices of environmental quality and sustainable development. In addition to publishing widely in these areas, he has participated in many review panels of the National Academy of Science, and consulted for the World Bank, the Asian Development Bank, the U.S. Agency for International Development, and the UN agencies specializing in natural resources and development; FAO, UNDP, UNEP and UNIDO. Professor Rogers has researched, consulted, and taught on these issues in over 30 countries. His recent books include, *America's Waters: Federal Roles and Responsibilities* (M.I.T. Press 1993) and *Water in the Arab World: Perspectives and Prognoses* (Harvard University Press 1994).

JEROME WIESNER was President Emeritus of the Massachusetts Institute of Technology until his death in 1994. He served as President of MIT from 1971–80 and upon his retirement as president resumed the title of Institute Professor, the Institute's highest faculty rank. Dr. Wiesner received a B.S. in Electrical Engineering and Mathematics and a Ph.D. in Electrical Engineering from the University of Michigan at Ann Arbor. Dr. Wiesner was an authority on microwave theory, communication science and engineering, signal processing, radio and radar propagation and phenomena, military technology, disarmament, and matters of science policy and technical education. During World War II he was a leader in the development of radar, and was later one of the principals in the conception and design of U.S. air defense and missile systems. Dr. Wiesner became a member of the President's

Science Advisory Committee in 1957. From 1961 to 1964, Dr. Wiesner was Special Assistant to President Kennedy for Science and Technology and, simultaneously, Chairman of the President's Science Advisory Committee. In 1964 Dr. Wiesner returned to MIT as dean of the MIT School of Science, and was appointed provost in 1966. Dr. Wiesner was a Fellow of the Institute of Electrical and Electronics Engineering, and of the American Academy of Arts and Sciences, as well as a Member of the National Academy of Sciences, the American Geophysical Union and numerous other institutions. He received numerous honorary degrees from various universities, and was the recipient of the President's Certificate of Merit and many other professional and public service awards. He has published extensively and is the author of the book, *Where Science and Politics Meet* (McGraw Hill 1961) and co-author of *ABM An Evaluation of the Decision to Deploy an Antiballistic Missile System* (Harper and Row 1969).